ISBN 978-1-330-26712-7
PIBN 10007058

English
Français
Deutsche
Italiano
Español
Português

www.forgottenbooks.com

Mythology Photography **Fiction**
Fishing Christianity **Art** Cooking
Essays Buddhism Freemasonry
Medicine **Biology** Music **Ancient
Egypt** Evolution Carpentry Physics
Dance Geology **Mathematics** Fitness
Shakespeare **Folklore** Yoga Marketing
Confidence Immortality Biographies
Poetry **Psychology** Witchcraft
Electronics Chemistry History **Law**
Accounting **Philosophy** Anthropology
Alchemy Drama Quantum Mechanics
Atheism Sexual Health **Ancient History**
Entrepreneurship Languages Sport
Paleontology Needlework Islam
Metaphysics Investment Archaeology
Parenting Statistics Criminology
Motivational

ENTIETH CENTURY TEXT-BOOKS

EDITED BY

A. F. NIGHTINGALE, Ph. D., LL.D.

SUPERINTENDENT OF SCHOOLS, COOK COUNTY, ILLINOIS

TWENTIETH CENTURY TEXT-BOOKS

ANIMAL STUDIES

A TEXT-BOOK OF ELEMENTARY ZOOLOGY
FOR USE IN HIGH SCHOOLS AND COLLEGES

BY

DAVID STARR JORDAN

PRESIDENT OF LELAND STANFORD JR. UNIVERSITY

VERNON LYMAN KELLOGG

PROFESSOR OF ENTOMOLOGY

AND

HAROLD HEATH

ASSOCIATE PROFESSOR OF INVERTEBRATE ZOOLOGY
IN LELAND STANFORD JR. UNIVERSITY

NEW YORK
D. APPLETON-CENTURY COMPANY
· INCORPORATED

EDITOR'S PREFACE

THE publishers are assured from the most expert testimony that they are presenting to the educational public, in this volume, a compact but complete treatment of elementary zoology, especially for those institutions of learning which prefer to find in a single book an ecological as well as a morphological survey of the animal world. Animal Life, in the series, treats largely of ecology, Animal Forms of morphology. In this volume the same authors present the salient features of each of the above-named books, with entirely new and additional chapters on Classification, the Economic Value, and the Ancient History of Animals. In Animal Studies an important advance step is taken toward meeting the demand for an elementary zoology that shall treat of the natural history rather than merely of the morphology of animals. Several new plates not found in the other volumes have been added. The structure, life habits, environments, economics, and history of fossil animals are each treated with that clearness, conciseness, and completeness for which President Jordan is particularly distinguished. While the book contains adequate material for a year's work, it is also admirably adapted for those schools that find it necessary to give a shorter time to this subject. The following ideas expressed by Dr. Coulter concerning

984125

the purpose of Plant Studies are entirely applicable to Animal Studies as well:

"The book is intended to serve as a supplement to three important factors:

"(1) The teacher, who must amplify and suggest at every point; (2) the laboratory, which must bring the pupil face to face with plants and their structures; (3) field-work, which must relate the facts observed in the laboratory to their actual place in Nature, and must bring new facts to notice which can be observed nowhere else."

Jordan and Price's Animal Structures, a Laboratory Manual of Zoology, will be found eminently useful in connection with the reading and study of this text.

CONTENTS

ANIMAL STUDIES

CHAPTER I

CONDITIONS OF ANIMAL LIFE

1. Divisions of the subject.—Biology is the science which treats of living things in all their relations. It is subdivided into Zoology, the science which deals with animals, and Botany, which is concerned with plants. The field covered by each of these branches is very extensive. Within the scope of zoology are included all subjects bearing on the form and structure of animals, on their development, and on their activities, including the consideration of their habits and the wider problems of their distribution and their relations to one another.

These various subjects are often conveniently grouped under three heads: Morphology, which treats of the form and structure or the anatomy of organisms; Physiology, which considers their activities; and Ecology, which includes their relations one to another and to their surroundings. All the phases of plant or animal existence may be considered under one or another of these three divisions.

2. Difference between animals and plants.—It is easy to distinguish between the animal and plant when a butterfly is fluttering about a blossoming cherry tree or a cow feeding in a field of clover. It is not so easy, if it is, indeed, possible, to say which is plant and which is animal when the simplest plants are compared with the simplest animals. It is almost impossible to so define animals as to distinguish all of them from all plants, or so to define plants as to distinguish all of them from all animals. While most animals have the power of locomotion, some, like the sponges and polyps and barnacles and numerous parasites, are fixed. While most plants are fixed, some of the low aquatic forms have the power of spontaneous loco-motion, and all plants have some power of motion, as espe-cially exemplified in the revolution of the apex of the growing stem and root, and the spiral twisting of tendrils, and in the sudden closing of the leaves of the sensitive plant when touched. Among the green or chlorophyll-bearing plants the food consists chiefly of inorganic sub-stances, especially of carbon which is taken from the car-bonic-acid gas in the atmosphere, and of water. But some green-leaved plants feed also in part on organic food. Such are the pitcher-plants and sun-dews, and Venus-fly-traps, which catch insects and use them for food nutrition. But there are many plants, the fungi, which are not green —that is, which do not possess chlorophyll, the substance on which seems to depend the power to make organic matter out of inorganic substances. These plants feed on organic matter as animals do. The cells of plants (in their young stages, at least) have a wall composed of a peculiar carbohydrate substance called cellulose, and this cellulose was for a long time believed not to occur in the body of animals. But now it is known that certain sea-squirts (Tunicata) possess cellulose. It is impossible to find any set of characteristics, or even any one characteristic, which is possessed only by plants or only by animals. But nearly all of the many-celled plants and animals may be easily distinguished by their general characteristics. The power of breaking up carbonic-acid gas into carbon and oxygen and assimilating the carbon thus obtained, the presence of chlorophyll, and the cell walls formed of cellulose, are char-acteristics constant in all typical plants. In addition, the fixed life of plants, and their general use of inorganic substances for food instead of organic, are characteristic$\!\!$ readily observed and practically characteristic of many celled plants. When the thousands of kinds of one-celled organisms are compared, however, it is often a matter of great difficulty or of real impossibility to say whether a given organism should be assigned to the plant kingdom or to the animal kingdom. In general the distinctive characters of plants are grouped around the loss of the power of locomotion and related to or dependent upon it.

3. **Living organic matter and inorganic matter.**—It would seem to be an easy matter to distinguish an organism—that is, a living animal or plant—from an inorganic substance. It is easy to distinguish a dove or a sunflower from stone, and practically there never is any difficulty in making such distinctions. But when we try to define living organic matter, and to describe those characteristics which are peculiar to it, which absolutely distinguish it from inorganic matter, we meet with some difficulties. At least many of the characteristics commonly ascribed to organisms, as peculiar to them, are not so. The possession of organs, or the composition of the body of distinct parts, each with a distinct function, but all working together, and depending on each other, is as true of a steam-engine as of a horse. That the work done by the steam-engine depends upon fuel is true; but so it is that the work done by the horse depends upon fuel, or food as we call it in the case of the animal. The oxidation or burning of this fuel in the engine is wholly comparable with the oxidation of the food, or the muscle and fat it is turned into, in the horse's body. The composition of the bodies of animals and plants of tiny structural units, the cells, is in many ways comparable with the composition of some rocks of tiny structural units, the crystals. But not to carry such rather quibbling comparisons too far, it may be said that organisms are distinguished from inorganic substances by the following characteristics: Organization; the power to make over inorganic substances into organic

matter, or the changing of organic matter of one kind, as plant matter, into another kind, as animal matter; motion, the power of spontaneous movement in response to stimuli; sensation, the power of being sensible of external stimuli; reproduction, the power of producing new beings like themselves; and adaptation, the power of responding to external conditions in a way useful to the organism. Through adaptation organisms continue to exist despite the changing of conditions. If the conditions surrounding an inorganic body change, even gradually, the inorganic body does not change to adapt itself to these conditions, but resists them until no longer able to do so, when it loses its identity.

4. **Primary conditions of animal life.**—Certain primary conditions are necessary for the existence of all animals. We know that fishes can not live very long out of water, and that birds can not live in water. These, however, are special conditions which depend on the special structure and habits of these two particular kinds of backboned animals. But the necessity of a constant and sufficient supply of air is a necessity common to both; it is one of the primary conditions of their life. All animals must have air. Similarly both fishes and birds, and all other animals as well, must have food. This is another one of the primary conditions of animal life. That backboned animals must find somehow a supply of salts or compounds of lime to form into bones is a special condition peculiar to these animals.

5. **Food.**—All the higher plants, those that are green (chlorophyll-bearing), can make their living substance out of inorganic matter alone—that is, use inorganic substances as food. But animals can not do this. They must have already formed organic matter for food. This organic matter may be the living or dead tissues of plants, or the living or dead tissues of animals. For the life of animals it is necessary that other organisms live, or have lived. It is this need which primarily distinguishes an animal from a

plant. Animals can not exist without plants. The plants furnish all animals with food, either directly or indirectly. The amount of food and the kinds of food required by various kinds of animals are special conditions depending on the size, the degree of activity, the structural character of the body, etc., of the animal in question. Those which do the most need most. Those with warmest blood, greatest activity, and most rapid change of tissues are most dependent on abundance, regularity, and fitness of their food. As we well know, an animal can live for a longer or shorter time without food. Men have fasted for a month, or even two months. Among cold-blooded animals, like the reptiles, the general habit of food taking is that of an occasioual gorging, succeeded by a long period of abstinence. Many of the lower animals can go without food for surprisingly long periods without loss of life. But the continued lack of food results inevitably in death. Any animal may be starved in time.

If water be held not to be included in the general conception of food, then special mention must be made of the necessity of water as one of the primary conditions of animal life. Protoplasm, the basis of life, is a fluid, although thick and viscous. To be fluid its components must be dissolved or suspended in water. In fact, all the truly living substance in an animal's body contains water. The water necessary for the animal may be derived from the other food, all of which contains water in greater or less quantity, or may be taken apart from the other food, by drinking or by absorption through the skin. Sheep are seldom seen to drink, for they find almost enough water in their green food. Fur seals never drink, for they absorb the water needed through pores in the skin.

6. **Oxygen.**—Animals must have air in order to live, but the essential element of the air which they need is its oxygen. For the metabolism of the body, for the chemical

changes which take place in the body of every living animal, a supply of oxygen is required. This oxygen is derived directly or indirectly from the air. The atmosphere of the earth is composed of 79.02 parts of nitrogen (including argon), .03 parts of carbonic acid, and 20.95 parts of oxygen. Thus all the animals which live on land are enveloped by a substance containing nearly 21 per cent of oxygen. But animals can live in an atmosphere containing much less oxygen. Certain mammals, experimented on, lived without difficulty in an atmosphere containing only 14 per cent of oxygen; when the oxygen was reduced to 7 per cent serious disturbances were caused in the animal's condition, and death by suffocation ensued when 3 per cent of oxygen was left in the atmosphere. Animals which live in water get their oxygen, not from the water itself (water being composed of hydrogen and oxygen), but from air which is mechanically mixed with the water. Fishes breathe the air which is mixed with or dissolved in the water. This scanty supply therefore constitutes their atmosphere, for in water from which all air is excluded no animal can breathe. Whatever the habits of life of the animal, whether it lives on the land, in the ground, or in the water, it must have oxygen or die.

7. **Temperature, pressure, and other conditions.**—Some physiologists include among the primary or essential general conditions of animal life such conditions as favorable temperature and favorable pressure. It is known from observation and experiment that animals die when a too low or a too high temperature prevails. The minimum or maximum of temperature between which limits an animal can live varies much among different kinds of animals. It is familiar knowledge that many kinds of animals can be frozen and yet not be killed. Insects and other small animals may lie frozen through a winter and resume active life again in the spring. An experimenter kept certain fish frozen in blocks of ice at a temperature of $-15°$ C.

for some time and then gradually thawed them out un-hurt. Only very hardy kinds adapted to the cold would, however, survive such treatment. There is no doubt that every part of the body, all of the living substance, of these fish was frozen, for specimens at this temperature could be broken and pounded up into fine ice powder. But a tem-perature of $-20°$ C. killed the fish. Frogs lived after being kept at a temperature of $-28°$ C., centipedes at $-50°$ C., and certain snails endured a temperature of $-120°$ C. without dying. At the other extreme, instances are known of ani-mals living in water (hot springs or water gradually heated with the organisms in it) of a temperature as high as $50°$ C. Experiments with *Amœbœ* show that these simplest animals contract and cease active motion at $35°$ C., but are not killed until a temperature of $40°$ to $45°$ C. is reached.

The pressure or weight of the atmosphere on the sur-face of the earth is nearly fifteen pounds on each square inch. This pressure is exerted equally in all directions, so that an object on the earth's surface sustains a pressure on each square inch of its surface exposed to the air of fifteen pounds. Thus all animals living on the earth's surface or near it, live under this pressure, and know no other condi-tion. For this reason they do not notice it. The animals that live in water, however, sustain a much greater pres-sure, this pressure increasing with the depth. Certain ocean fishes live habitually at great depths, as two to five miles, where the pressure is equivalent to that of many hundred atmospheres. If these fishes are brought to the surface their eyes bulge out fearfully, being pushed out through reduced expansion; their scales fall off because of the great expansion of the skin, and the stomach is pushed out from the mouth till it is wrong side out. Indeed, the bodies sometimes burst. Their bodies are accustomed to this great pressure, and when this outside pressure is sud-denly removed the body may be bursted. Sometimes such a fish is raised from its proper level by a struggle with its

prey, when both captor and victim may be destroyed by
the expansion of the body.   Some fishes die on being taken
out of water through the swelling of the air bladder and
the bursting of its blood-vessels.   If an animal which lives
normally on the surface of the earth is taken up a very high
mountain or is carried up in a balloon to a great altitude
where the pressure of the atmosphere is much less than it
is at the earth's surface, serious consequences may ensue,
and if too high an altitude is reached death occurs.   This
death may be in part due to the difficulty in breathing in
sufficient oxygen to maintain life, but it is probably chiefly
due to the disturbances caused by the removal of the pressure to which the body is accustomed and is structurally
adapted to withstand.   All living animals are accustomed
to live under a certain pressure, and there are evidently
limits of maximum or minimum pressure beyond which no
animal at present existing can go and remain alive.

But in the case both of temperature and pressure conditions it is easy to conceive that animals might exist which
could live under temperature and pressure conditions not
included between the minimum and maximum limits of each
as determined by animals so existing.   But it is impossible
to conceive of animals which could live without oxygen or
without organic food.   The necessities of oxygen and organic
food (and water) are the primary or essential conditions for
the existence of any animals.

Of course, we might include such conditions, among
the primary conditions, as the light and heat of the sun,
the action of gravitation, and other physical conditions,
without which existence or life of any kind would be impossible on this earth.   But we here consider by " primary
conditions of animal life " rather those necessities of
living animals as opposed to the necessities of living
plants.   Neither animals nor plants could exist without
the sun, whence they derive directly or indirectly all their
energy.

8. **Cells.**—If we examine very carefully the different parts of some highly developed animal under the high powers of the microscope we find that they are composed of a multitude of small structures which bear the same relations to the various organs that bricks or stones do to a wall; and if the investigation were continued it would be found that every organism is composed of one or more of these lesser elements which bear the name of *cells*. In size they vary exceedingly, and their shapes are most diverse, but, despite these differences, it will be seen that all exhibit a certain general resemblance one to the other.

9. **Shape of cells.**—In many of the simpler organisms the component cells are jelly-like masses of a more or less spherical form, but as we ascend the scale of life the condition of affairs becomes much more complex. In the muscles the cells are long and slender (Fig. 1, D); those forming the nerves and conveying sensations to and from all parts of the body, like an extensive telegraph system, are excessively delicate and thread-like; in the skin, and lining many cavities of the body, where the cells are united into extensive sheets, they range in shape from high and columnar to flat and scale-like forms (Fig. 1, E, F, G). The cells of the blood present another type (Fig. 1, B); and so we might pass in review other parts of the body, and continue our studies with other groups of animals, always finding new forms dependent upon the part they play in the organism.

10. **Size of cells.**—Also in the matter of size the greatest variations exist. Some of the smallest cells measure less than one micromillimeter ($\frac{1}{25000}$ of an inch) in diameter. Over five hundred million such bodies could be readily stowed away into a hollow sphere the size of the letter beginning this sentence. In a drop of human blood of the same size, between four and five million blood-cells or corpuscles float. And from this extreme all sizes exist up to those with a diameter of 2.5 or 5 c.m. (one or two inches),

as in the case of the hen's or ostrich's egg. On the average a cell will measure between .025 to .031 mm. ($\frac{1}{1000}$ and $\frac{1}{800}$ of an inch) in diameter, a speck probably invisible to the unaided eye. While the size and external appearance of a cell are seen to be most variable, the internal structures are found to show a striking resemblance throughout. All are constructed upon essentially the same plan. Differences in form and size are superficial, and in passing to a more careful study of one cell we gain a knowledge of the important features of all.

11. **A typical cell.**—An egg-cell (Fig. 1, A) or some simple one from the liver or skin may be chosen as a good representative of a typical cell. To the naked eye it is barely visible as a minute speck; but under the microscope the appearance is that of so much white of egg, an almost transparent jelly-like mass bearing upon its outer surface a thin structureless membrane that serves to preserve its general shape and also to protect the delicate cell material within. The comparison of the latter substance to egg albumen can be carried no further than the simple physical appearance, for albumen belongs to that great class of substances which are said to be non-living or dead, while the cell material or *protoplasm*, as it is termed, is a living substance. We know of no case where life exists apart from protoplasm, and for this reason the latter is frequently termed the physical basis of life.

In addition to the features already described, the protoplasm of every perfect cell is modified upon the interior to form a well-defined spherical mass known as the *nucleus*. Other structures are known to occur in the typical cell. Experiment shows that the nucleus and cell protoplasm are absolutely indispensable, whatever their size and shape, and therefore we are at present justified in defining the cell as a small mass of protoplasm enclosing a nucleus.

12. **Structure of protoplasm.**—When seen under a glass of moderate power protoplasm gives no indication of any

definite structure, and even with the highest magnification it presents appearances which are not clearly understood. According to the commonly accepted view, it consists of two portions, one, the firmer, forming an excessively delicate

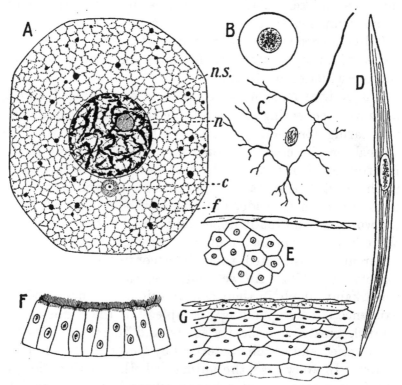

Fig. 1.—Different types of cells composing the body of a highly developed animal. A, cell; *f*, food materials; *n*, nucleus, B, blood-cell. C, nerve-cell with small part of its fiber. D, muscle fiber. E, cells lining the body cavity. F, lining of the windpipe. G, section through the skin. Highly magnified.

meshwork (Fig. 1, A) enclosing in its cavities the second more fluid part. Therefore, when highly magnified, the appearance would be essentially like a sponge fully saturated with water; but it should be remembered that in the protoplasm the sponge work, and possibly the fluid part, is living, and that both are transparent.

There are reasons for thinking that the structure and.

the composition of protoplasm may change somewhat under certain circumstances. It certainly is not everywhere alike, for that of one animal must differ from that of another, and different parts, such as the liver and brain, of the same form must be unlike. These differences, however, are minor when compared to the resemblances, for, as we shall see, this living substance, wherever it exists, carries on the processes of waste, repair, growth, sensation, contraction, and the reproduction of its kind.

13. **Animal functions.**—Animals in general lead active, busy lives, collecting food, avoiding enemies, and producing and caring for their young. While the activities of all animals are directed to their own preservation and to the multiplication of their kind, these processes are carried on in the most diverse ways. The manner in which an organ or an organism is made, and the method by which it does its work, are mutually dependent one on the other. As there is an enormous number of species of animals, each differently constructed, there is, accordingly, a very great variety of habits. As we shall see, the lower forms are remarkably simple in their construction, and their mode of existence is correspondingly simple. In the higher types a much greater complexity exists, and their activities are more varied and are characterized by a high degree of elaboration. In every case, the animal, whether high or low, is fitted for some particular haunt, where it may perform its work in its own special way and may lead a successful life of its own characteristic type.

CHAPTER II

14. Classification.—It is plain that natural relations of some sort exist among living organisms. A dog is more like a cat than it is like a sheep. A dog is more like a sheep than either is like a butterfly. The very existence of such terms as animals and plants, insects and fishes, implies various grades of relationship. Classification is the process of reducing our knowledge of these grades of likeness and unlikeness to a system. By bringing together those which are alike, and separating those which are unlike, we find that these rest on fixed and inevitable laws. Classification is thus defined as "the rational, lawful disposition of observed facts."

15. Homology.—All rational classification of plants or animals concerns itself with homologies. Homology means fundamental identity of structure, as distinguished from analogy, which means incidental resemblance in form or function. Thus the arm of a man is homologous with the fore leg of a dog, because in either we can trace throughout deep-seated resemblances or homologies with the other. In every bone, muscle, vein, or nerve the one corresponds closely with the other. The "limb" of a tree, the "arm" of a starfish, or the fore leg of a grasshopper shows no such correspondence. In a natural classification, or one founded on fact, those organisms showing closest homologies are placed together. An artificial classification is one based on analogies. Such a classification would place together a

13

cricket, a frog, and a kangaroo, because they all jump; or a bird, a bat, and a butterfly, because they all have wings and can fly, although the different kinds of wings are made in very unlike fashion.

16. **Natural classification based on homology.**—The closest homologies are shown by those animals which have sprung from a common stock. The basis of natural classification, which is an expression of the ancestry of blood relationship of animals, is therefore homology. So far as we know, the actual presence of homologies among animals implies their common descent from some stock possessing the same characters. The close resemblance or homology among the different races of men indicates that all men originally came from one stock. As homology implies blood-relationship, so, on the other hand, common descent implies homology, the similar parts being derived from a common ancestral stock. It is sometimes said that the inside of an animal tells what it is, the outside where it has been. In the internal structure, ancestral traits are perpetuated with little change through long periods. The ex-

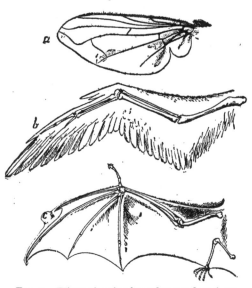

Fig. 2.—Wings showing homology and analogy. *a*, fly; *b*, bird; *c*, bat.

ternal characters, having more to do with surroundings, are much more rapidly altered in response to demands of the environment.

A perfect classification would indicate the line of descent of each member of the series, those now living having sprung in natural sequence, by slow processes of change, from creatures of earlier geological periods. It is said that in classification we have " three ancestral documents": Morphology, Embryology, and Paleontology. In Morphology we compare one form with another, thus

a. Elephant.   *b.* Coney.   *c.* Rhinoceros.   *d.* Horse.

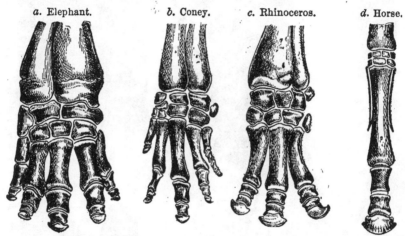

Fɪɢ. 3.—Homology of digits of four odd-toed mammals, showing gradual reduction in number and consolidation of bones above.—After Roᴍᴀɴᴇs.

tracing resemblances and differences. In Embryology we trace the development of individuals from the egg, thus finding clues in heredity that will enable us to trace the development of the race. In Paleontology we study the extinct forms directly, thus often finding evidence as to the origin of forms now existing.

17. **Scientific names.**—Each of the different kinds of animal or plant is called a species. There is no better definition of species. Thus the red squirrel is a kind or species of squirrel, the gray squirrel is another, the fox squirrel a third. The black squirrel of the East is not a species, because black squirrels and gray squirrels are sometimes found in the same nest, born from the same parents.

A genus is a group of closely related species—one or more—separated from other genera by tangible structural characters. Thus all the squirrels named above constitute

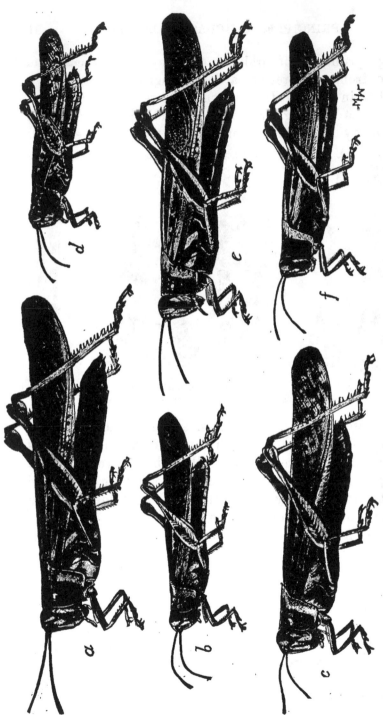

FIG. 4.—Locusts (*Schistocerca*) taken on the Galapagos Islands, Pacific Ocean: all descended from a common ancestor, but now scattered over the various islands, and varying in size and markings. *a*, *Schistocerca melanocera* Stol (Charles Island); *b*, *S. intermedia borealis* Snodgrass (Abingdon and Bindloe Islands); *c*, *S. intermedia* Snodgrass (Duncan Island); *d*, *S. literosa* Walker (Chatham Island); *e*, *S. melanocera lineata* Snodgrass (Albemarle Islands); *f*, *S. melanocera immaculata* Snodgrass (Indefatigable Island). (The species *intermedia* is probably a hybrid between the two other species.)

a single genus. Other squirrel-like animals, as the chipmunk, the flying squirrel, the prairie gopher, or the prairie dog, belong to as many different genera.

In the binomial system, invented by Linnæus and applied by him to animals in 1758, the scientific name of an animal consists of two words—the name of its genus and species taken together. The name of the genus comes first. It is a noun, in Latin form, though usually of Greek derivation—thus *Sciurus,* the squirrel, in Greek meaning shadow-tail. The name of the species is an adjective in meaning, placed after the noun and agreeing with it. Thus *Sciurus hudsonicus* is the name of the red squirrel, *Sciurus carolinensis* of the Eastern gray squirrel, and *Sciurus ludovicianus* of the fox squirrel; *Sciuropterus volans* is the flying squirrel, *Tamias striatus* the Eastern chipmunk, and *Spermophilus franklini* one of the prairie gophers. The specific name is usually a descriptive adjective—often the name of a locality, sometimes the name of a man. The authority usually written after the name of an animal is that of the one who gave it its specific name —thus *Sciurus hudsonicus* Erxleben, which means that Erxleben first called it *hudsonicus.* Usually the name of the authority is that of the discoverer of the species. When several names are given to the same animal they are called synonyms. The earliest of these names is the right name. All the rest are wrong.

18. **Families of animals.**—A group of related genera is called a family. The name of a family is derived from that of its principal genus, with the termination *idæ.* Thus all the squirrel-like animals belong to the family of *Sciuridæ.* All the sorts of mice to the *Muridæ*, from the principal genus, *Mus*, the mouse. The rabbits are *Leporidæ*, from *Lepus*, the rabbit, and the beavers *Castoridæ*, from *Castor*, the beaver.

19. **Higher groups of animals.**—In the higher groups we first trace out the different plans of structure. There is

Fɪɢ. 5.—Three species of jack-rabbits, differing in size, color, and markings, but
believed to be derived from a common stock. The differences have arisen
through isolation and adaptation. The upper figure shows the head and fore legs
of the black jack-rabbit (*Lepus insularis*), of Espiritu Santo Island, Gulf of
California ; the lower right-hand figure, the Arizona jack-rabbit (*Lepus alleni*),
specimen from Fort Lowell, Arizona ; and the lower left-hand figure is the San
Pedro Martir jack-rabbit (*Lepus martirensis*), from San Pedro Martir, Baja
California.

some question as to the number of these different types, but we are not far out of the way in recognizing seven principal ones. These give rise to the seven principal branches of the animal kingdom: Protozoa, Cœlenterata, Mollusca, Vermes, Arthropoda, Echinodermata, and Chordata (which includes vertebrates). The followers of Cuvier and Agassiz reduced these to four or five: Protozoa, Radiata, Mollusca, Articulata, and Vertebrata; but a more thorough knowledge of the different groups makes the larger number preferable, the radiates and the articulates being each divided into two. Many zoologists break up the Vermes into several distinct branches.

The branches are again divided into classes. Thus the mammals, birds, reptiles, amphibians, fishes, lampreys, and lancelets are classes of vertebrates. The insects form a class of Arthropods.

Each class is again divided into orders. The Glires or rodents, the gnawing animals, of which squirrels, mice, and rabbits are examples, form an order of mammals. The hoofed animals, Ungulata, form another, and each of these again contains many families.

Intermediate divisions are sometimes recognized, with the prefixes *super* and *sub*. A subfamily is a division of a family including certain genera. A superfamily is a group of related families within the limits of an order.

The red squirrel belongs to the branch Chordata, class Mammalia, order Glires, family Sciuridæ, genus *Sciurus*, species *Hudsonicus*.

20. **Trinomial names.**—Trinomial names are those in which the binomial name of a species is followed by a second adjective. These indicate subspecies or varieties connected with geographical distribution. Thus many forms have a northern variety, a southern variety, one in the mountains, one on the plains, in the forests, or in other peculiar situations.

Thus the gray squirrel, typically southern, has a sub-

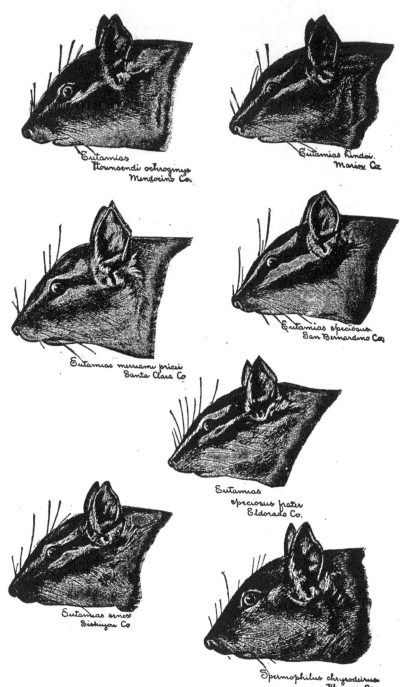

Eutamias
Townsendi ochroginus
Mendocino Co.

Eutamias hindsi.
Marin Co.

Eutamias merriami pricii
Santa Clara Co

Eutamias speciosus
San Bernardino Co.

Eutamias
speciosus frater
Eldorado Co.

Eutamias senex
Siskiyou Co

Spermophilus chrysodeirus
Plumas Co

FIG. 6.—Some chipmunks of California, showing distinct species produced through isolation.—From nature, by WILLIAM SACKSTON ATKINSON.

species, *Sciurus carolinensis leucotis* (white-eared), in the Northern States, larger than the true *Sciurus carolinensis*, with the dark band on the back narrower. In Minnesota is another subspecies, *Sciurus carolinensis hypophæus*, with only a narrow streak of white on the belly. As animals come to be better known we can recognize by name more and more of these subspecies or geographical variations.

Even in the same locality the members of a species vary more or less, no two being exactly alike. The name variety is applied to any sort of variation which can be recognized. Usually varieties not having definite geographical range receive no scientific name. When forms in different geographical areas are found to intergrade or mix with one another they are known as subspecies, the one first named being regarded as the original species. When they do not intergrade they are called distinct species. The subspecies differ from the species in degree only. When the range of a species is crossed by an impassable barrier, the subspecies on either side of the barrier usually becomes in time a distinct species. Thus distinct species are said to be produced through isolation. The plates which follow may serve as illustrations of species and subspecies thus formed.

CHAPTER III

21. Single-celled and many-celled animals.—In almost every portion of the globe there are multitudes of animals whose body consists of but a single cell; while those forms more familiar to us, and usually of comparatively large size and higher development, such as sponges, insects, fishes, birds, and man himself, are composed of a multitude of cells. For this reason the animal kingdom has been divided into two great subdivisions, the Protozoa including all unicellular forms and the Metazoa embracing those of many cells.

22. Single-celled animals.—The division of the Protozoa comprises a host of animals, usually of microscopic size, inhabiting fresh or salt water or damp localities on land in nearly every portion of the globe. The greater number wage their little, though fierce, wars on one another without attracting much attention; others, in the sharp struggle, have been compelled to live upon or within the bodies of other animals, and many have become notorious because of the diseases they produce under such circumstances. A few are in large measure responsible for the phosphorescence of the sea; and still others have long been favorite objects of study because of their marvelous beauty. Adapted for living under diverse conditions, the bodily form differs greatly, and yet all conform to three or four principal types, of which we may gain a good idea from the study of a few representative forms.

22

23. The Amœba.—Among the simplest one-celled animals living in the ooze at the bottom of nearly every freshwater stream or pond is the *Amœba* (Fig. 7, A), whose body is barely visible to the unaided eye.   Under the microscope

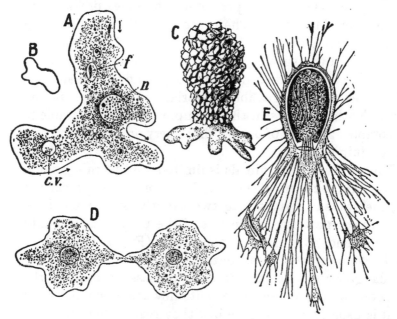

Fig. 7.—A, the *Amœba*, highly magnified, showing *c. v.*, pulsating vacuole; *f*, food particle; *n*, nucleus. The arrows show the direction of movement. B, shape of same individual 30 seconds later. C, an amœba-like animal (*Difflugia*) partially enclosed in a shell. D, an *Amœba* in the process of division. E, Gromia, another shelled protozoan.—After SCHULZE.

it is seen to consist of an irregular, jelly-like mass of protoplasm totally destitute of a cell wall.   Unlike those animals with which we are familiar, the body constantly changes its shape.   A rounded bud-like projection will be seen to appear on one side of the body and the protoplasm of adjacent regions flows into it, thereby increasing its extent.   Similar projections at the opposite end of the cell are withdrawn, and their substance may flow into the newly formed lobe, which gradually swells in size and pushes forward.   Thus, by constantly advancing the front part of the body and retracting the hinder portion, the cell glides or flows along from place to place.

Upon meeting with any of the smaller organisms upon which it lives, projections from the body are put out which gradually flow around the prey and it becomes pressed into the interior of the cell. The process is not unlike pushing a grain of sand into a bit of jelly. There is no mouth. Any point on the surface serves for the reception of food. Oxygen gas also is taken into the body all over the surface, and wastes and indigestible material are cast out at any point. Nothing exists in these simple forms comparable to the complex systems of organs that carry on these processes in the squirrel.

The bodily size of animals is limited, and to this general rule the *Amœba* is no exception, for upon gaining a certain size, the nucleus divides into two exactly similar portions, and very soon afterward the rest of the body separates into two independent masses of equal size (Fig. 7, D), each of which, when entirely free, contains a nucleus. In this way two daughter amoebæ are formed possessing exactly the characters of the parent save that they are of smaller size; but it is usually not long before they reach their limit of growth, when division occurs again, and so on, generation after generation.

It not infrequently happens, however, that the pond or stream, in which the *Amœba* and other Protozoa live, dries up for a portion of the year. In such an event the body assumes a spherical shape, develops a firm, horn-like membrane about itself, and thus *encysted* it withstands the summer's heat and dryness and may be transported by the wind, or otherwise, over great distances. When the conditions again become favorable the wall ruptures and the *Amœba* emerges to repeat its life processes.

24. Some relatives of the Amœba.—All amœba-like forms, to the number of perhaps a thousand species, possess this same method of locomotion, but many present some inter-

esting additional characters.   For example, the form represented in Fig. 7, C, constructs a sac-like skeleton of tiny
pebbles cemented together, into which it may withdraw for
protection.   Others construct similar envelopes of lime or
flint, and still others, as they continue to grow, build on
additional chambers, giving rise to a great variety of forms
often of wonderful beauty.   In the tropics, particularly,
some of the shelled Protozoa are so abundant that they may
impart a whitish tinge to the water, and in some places
their empty shells on falling to the bottom form immense
deposits.   The chalk cliffs of England are in large measure
made up of such shells.

25. **The Infusoria.**—A little over two hundred years ago
it was discovered that wherever water remained stagnant it
became favorable for the rapid multiplication of a large
number of species of Protozoa which live in such situations.
These are known as Infusoria, and, like the preceding species, are usually of microscopic size and of the most varied
shapes.   The first striking feature of their organization is
the presence of a delicate though relatively firm external
cell membrane known as the *cuticle*, which preserves a definite shape to the body.   Such a method of locomotion as
exists in the preceding group is consequently an impossibility, but other and more highly developed structures perform the office.   These latter organs are of two types, and
their general characteristics may be readily understood
from an examination of a few species living in the same
localities as the *Amœba*.

26. **The Euglena.**—The first type exists in the common
fresh-water organism known as *Euglena*, represented in
Fig. 8, A.   Here the spindle-shaped body is surrounded by
a delicate cuticle perforated at one point, where a funnel-
shaped depression, the gullet, leads into the soft protoplasmic interior.   From the base of this depression the
protoplasm is drawn out in the form of a delicate whip-like
process known as the *flagellum*.   This structure, always

permanent in form, constantly beats backward and forward with great rapidity in a general direction represented in the diagram (Fig. 8, *c*). The movement from *a* to *b* is much more rapid than the reverse, from *b* to *a*, which results, like the action of the human arm in swimming, in driving the organism forward. Not only does the flagellum serve the purpose of locomotion, but it also produces currents in the water which may serve to bear minute organisms down into the gullet, whence they readily pass into the soft pro-

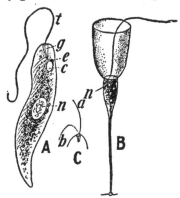

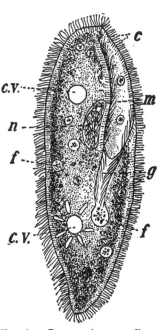

Fig. 8. — Flagellate Infusoria. **A,** *Euglena viridis* ; *c,* pulsating vacuole ; *e,* eye-spot ; *g,* gullet ; *n,* nucleus ; *t.* flagellum. B, *Codosiga,* with collar surrounding the flagellum. C, diagram illustrating the action of the flagellum. All figures greatly enlarged.

Fig. 9. — *Paramœcium aurelia,* a ciliate infusorian. *c,* cilia; *c.v.,* pulsating vacuoles ; *f,* food particles ; *g,* gullet ; *m,* buccal groove ; *n,* nucleus.

toplasm of the body, there to undergo the processes of digestion and assimilation. In some forms the protoplasm in the region of the flagellum is drawn out in the form of a collar (Fig. 8, B), whose vibratory motion also aids in conveying and guiding food into the body.

27. **The Slipper Animalcule.**—The second type of locomotor organ may be understood from a study of the

Slipper Animalcule (*Paramœcium*, Fig. 9), abundant in stagnant water. In this form the cuticle surrounding the somewhat cylindrical body is perforated by a great number of minute openings through which the internal protoplasm projects in the form of delicate threads. Each process, termed a *cilium*, works on the same principle as the flagellum, but it beats with an almost perfect rhythm and in unison with its fellows, drives the animal hither and thither with considerable rapidity.

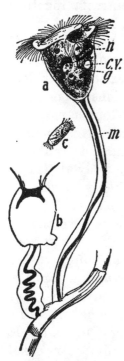

FIG. 10.—*Vorticella*, an attached ciliate infusorian, highly magnified. a, fully extended individual; *c.v.*, pulsating vacuole; *g*, gullet; *n*, nucleus. b, contracted specimen. c, small free-swimming individual, which unites with a stationary individual (one partly united is shown in specimen b).

On one side of the body is a furrow which deepens as it runs backward and finally passes into the gullet (*g*), which leads into the interior of the body. Throughout the entire extent it is lined with cilia which create strong currents in the surrounding water and in this way conduct food down the gullet into the body. Embedded in the outer surface of the body, in among the cilia, are also a number of very minute sacks, each containing a coiled thread which may be discharged against the body of any intruder, so that this form is supplied with actual organs of defense. Two *pulsating vacuoles* (*c.v.*) or simple kidneys are also present, consisting of a central reservoir into which a number of radiating canals extend.

28. **The Bell Animalcule and other species.**—The Bell Animalcule (*Vorticella*, Fig. 10) is often found in the same situations as the Slipper Animalcule, which in certain respects it resembles. It is generally attached by a slender stalk, and where many

are growing together they appear like a delicate growth of mold upon the water weed. The stalk is peculiar in being traversed by a muscle fiber arranged in a loose spiral, which, upon any unusual disturbance, contracts together with the body into the form shown in Fig. 10, *b*.

These few examples serve to show the general plan of organization and the method of locomotion of the Infusoria; but, as upward of a thousand species exist, with widely differing habits, many interesting modifications are present. Some have been driven in past time to adopt a parasitic mode of life within the bodies of other animals. At present they are devoid of locomotor organs, and as they absorb nutritive fluids through the surface of the body all traces of a mouth áre also absent. The reproductive processes also are peculiar, but they do not concern us now.

29. **Gregarina.**—Another type of protozoan worthy of special attention is that of the *Gregarina* (Fig. 11), various species of which live in the alimentary canal * of crayfishes and centipeds and certain insects. *Gregarina* is a parasite, living at the expense of the host in whose body it lies. It has no need to swim about quickly, and hence has no swimming cilia like *Paramœcium* and the young *Vorticella*. It does need to cling to the inner wall of the alimentary canal of its host, and the body of some species is provided with hooks for that purpose. The food of *Gregarina* is the liquid food of the host as it exists in the intestine, and which is simply absorbed anywhere through the surface of the body of the parasite. There is no mouth opening nor

---

* Specimens of *Gregarina* can be abundantly found in the alimentary canal of meal worms, the larvæ of the black beetle (*Tenebrio molitor*), common in granaries, mills, and brans. "Snip off with small scissors both ends of a larva, seize the protruding (white) intestine with forceps, draw it out, and tease a portion in normal salt solution (water will do) on a slide. Cover, find with the low power (minute, oblong, transparent bodies), and study with any higher objective to suit."— MURBACH.

fixed point of ejection of waste material, nor is there any
contractile vacuole in the body.

In the method of multiplication or reproduction *Gregarina* shows an interesting difference from *Amœba* and
Paramœcium and *Vorticella*.   When the *Gregarina* is

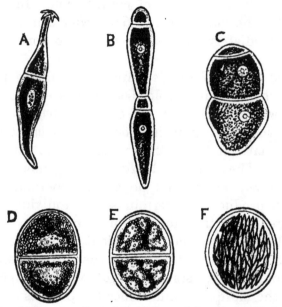

Fig. 11.—Gregarinidæ.   A, a Gregarinid (*Actinocephalus oligacanthus*) from the intestine of an insect (after STEIN) ; B and C, spore forming by a Gregarinid (*Coccidium oviforme*) from the liver of a guinea-pig (after LEUCKART) ; D, E, and F,
successive stages in the conjugation and spore forming of *Gregarina polymorpha*
(after KOELLIKER).

ready to multiply, its body, which in most species is rather
elongate and flattened, contracts into a ball-shaped mass
and becomes encysted—that is, becomes inclosed in a tough,
membranous coat.   This may in turn be covered externally
by a jelly-like substance.   The nucleus and the protoplasm
of the body inside of the coat now divide into many small
parts called spores; each spore consisting of a bit of the
cytoplasm inclosing a small part of the original nucleus.
Later, the tough outer wall of the cyst breaks, and the
spores fall out, each to grow and develop into a new *Gre-*

garina. In some species there are fine ducts or canals leading from the center of the cyst through the wall to the outside, and through these canals the spores issue. Sometimes two *Gregarinæ* come together before encystation and become inclosed in a common wall, the two thus forming a single cyst. This is a kind of conjugation. In some species each of the young or new *Gregarinæ* coming from the spores immediately divides by fission to form two individuals.

Related to the *Gregarinæ* are those minute protozoan parasites which live in the blood-corpuscles of man and some of the lower animals, and are called *Hæmatozoa.* Three species of these, living in the blood of man, cause the three kinds of malarial fever, known as tertian, quartan, and remittent. These malarial *Hæmatozoa,* known generally as *Hæmamœba,* can multiply by asexual sporulation in the blood, but produce also certain sexual individuals, which, when taken into the stomach of a mosquito which has sucked blood from a malarial patient, give rise to a zygote which encysts in the outer walls of the stomach, and breaks up into numerous blasts or embryos, which escape into the blood of the mosquito, and thence to all parts of its body, and especially to the salivary or poison glands. When now this infected mosquito pierces the skin of another man, and pours into the wound, as it regularly does, a quantity of saliva, numbers of larval *Hæmamœbæ* also enter the blood, and, multiplying here, soon set up the disease malaria in the bitten person. It has been definitely proved that malaria is thus disseminated by mosquitoes, and it is highly probable that it is contracted in no other way.

30. **Characteristics common to the Protozoa.**—We have now studied the principal structures which serve in locomotion among these simple one-celled forms, also the means by which they catch their food, and we shall now glance at the internal processes, which are much the same in all.

After the food has been taken into the cell, it is probably acted upon by some digestive fluid, for it soon assumes a granular appearance, and finally undergoes complete solution.   In every case the oxygen is absorbed through the general surface of the body, and uniting with the living substance, as in the squirrel, liberates the energy necessary for the performance of the animal's life-work.   The wastes thus produced in a large number of forms simply filter out from the body without the agency of anything comparable to a kidney, but in several species they are borne to a definite spot, the pulsating vacuole (Figs. 7, 9, 10, *c.v.*), where they gradually accumulate into a drop about the size of the nucleus.   The wall between it and the exterior now gives way, and the excretions are passed out.   In active individuals this process may be repeated two or three times a minute, but it is usually of less frequent occurrence.

The loss in bodily waste is continually made good by the manufacture of the food into protoplasm, and if the income be greater than the outgo, growth ensues.   But, as in all other forms, growth is limited, and ultimately the cell is destined to divide, resulting in two new individuals.   This process may be repeated many times, but not indefinitely, for sooner or later various members of the same species unite in pairs temporarily or permanently, exchange nuclear material, and separate again with apparently renewed energy and the ability to divide for many generations.

31. **Simple and complex animals.**—It is important to note that these same processes of waste, repair, growth, feeling, motion, and multiplication are the same as those of the squirrel, and, furthermore, are common to all living creatures, so that the difference between animals is not in their activities, but in their bodily mechanisms; and according to the perfection of this, the animal is high or low in the scale.   Comparing, for example, the *Amœba* and Slipper Animalcule, which are relatively low and high Protozoa, we find in the former that any part of the body serves in locomotion and in the capture of food, while in the latter these

same functions are performed by definite structures, the cilia and gullet. Now, it is well known that a workman is able to make better watch-springs, when this is his sole duty, than another who must make all parts of the watch; and likewise, where a definite task is performed by a definite structure, it is more efficiently done than where any and every part of the body must carry it on. So the *Amœba*, in which definite tasks are performed by any part of the body indifferently, is less perfect and thus lower than the *Paramœcium*, where these functions are performed by special organs. As we ascend the scale of life we find this division of labor among special parts of the body more complete, the organs, and therefore the animal, more complex, and better fitted to carry on the work of its life.

CHAPTER IV

THE SLIGHTLY COMPLEX ANIMALS OR SPONGES

32. Their relation to the Protozoa.—While the greater number of one-celled forms are not united with their fellows, there are several species where the reverse is true. In Fig. 12, for example, a fresh-water form known as *Pandorina* is represented, consisting of sixteen cells embedded in a spherical, jelly-like substance, each one of which is precisely like its companions in form and activity. The aggregation may be looked upon as a colony of sixteen Protozoa united together to derive the benefit of increased locomotion and a larger amount of food in consequence. As a result of such a union they have not lost their independence, for if one be separated from the main company it continues to exist.

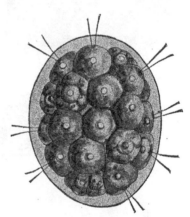

FIG. 12.—*Pandorina* (from Nature). Highly magnified.

From such a simple colonial type we may pass through a series of several more complex forms which reach their highest development in the beautiful organism, *Volvox* (Fig. 13). In this form the individual members, to the number of many thousand, are arranged in the shape of a hollow sphere. The united efforts of the greater number, which bear on their outer surfaces two flagella, drive the colony with the rolling movement

33

from place to place. As just indicated, some individuals lack the flagella, and their subsequent careers show them to be of a peculiar type. Sooner or later each· undergoes a series of divisions forming a little globe of cells, which migrates into the interior of the parent sphere and develops into a new colony. Within a short time the walls of the parent break, liberating the imprisoned young, which continue the existence of the species while the parent organism soon decays.

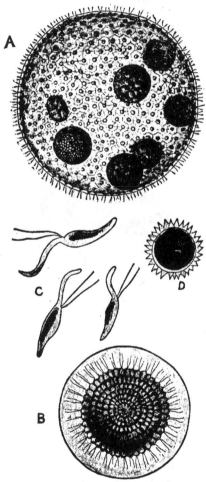

Under certain circumstances, instead of developing colonies by such a method, some of the cells may store up food matters and become eggs, while others, known as sperm-cells, develop a flagellum, and separating from the colony swim actively in the surrounding water, where each finally unites with an egg. This union, like that of the two individuals in *Vorticella* (Fig. 10, *b*, *c*), results in the power of division, and the egg enters upon its development, dividing again and again. The cells so produced remain together, form a sphere, and finally develop a *Volvox* colony.

FIG. 13.—A, *Volvox minor*, entire colony (from Nature). B, C, and D, reproductive cells of *Volvox globator*. All highly magnified.

In such associations as *Volvox* an important step has been taken beyond that of *Pandorina*, for there is a division of the labors of the colony among its various members, some acting as locomotor cells while others are germ-cells. These are now so dependent one upon the other that they are unable to exist after separation from the main company, just as a part of the squirrel is incapable of leading an independent existence. A higher type of organism has thus arisen intermediate between the simple one-celled animals and those of many cells, especially the sponges—a relation which is more readily recognized after an examination of the latter.

33. **Development of the sponge.**—Like all many-celled animals, the sponge *begins* its life, however, as a single cell —the egg—which is in this case barely visible to the sharp unaided eye. Fertilized by its union with a sperm cell, development commences, and the first apparent indication of the process will be the division of the cell into halves (Fig. 14, A, B). Each half redivides into four, these again into eight cells, and this process is repeated, giving the young sponge the general form of *Pandorina*. The divisions of the cells still continue and result in the formation of a hollow globe of cells (called the *blastula*, Fig. 14, E, F) similar to *Volvox*, and at this point the young larva leaves the parent.

The next transformation consists in a pushing in of one side of the sphere, just as one might press in the side of a hollow rubber ball. The depression gradually deepens, and finally results in the formation of a two-layered sac known as the *gastrula* (Fig. 14, G). At this stage of its existence the sponge settles down for life in some suitable spot, by applying the opening of its sac-like body to some foreign object. In assuming the final form a new mouth breaks through what was once the bottom of the sac, canals perforate the body wall, a skeleton is developed, and the characteristic features of the adult are thus attained.

34. Distribution.—The sponges are aquatic animals, and, with the exception of one family consisting of relatively few species, all are inhabitants of the sea in every part of

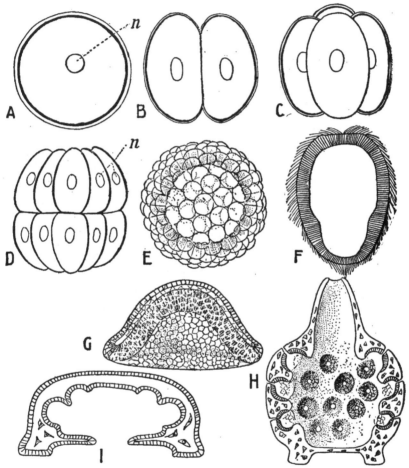

FIG. 14.—Diagrams illustrating the development of a sponge. A, egg-cell; *n*, nucleus. B, C, D, 2-, 4-, and 16-cell stages. E, *blastula*. F, section through somewhat older larva. G, *gastrula*. H, young sponge. I, section through somewhat younger larva than H.

the globe. The larger number occupy positions along the shore, becoming especially abundant in the tropics; but other species occur at greater depths, several species living

between three and four miles from the surface. Unlike
the majority of animals, all members of this group are
securely fastened to some foreign object, such as rocks, the
supports of wharves, or with one extremity embedded in
the sand. As we have seen, the young enjoy a free-swimming existence and are swept far and wide by means of
tidal currents, but sooner or later these migrations are
terminated in some suitable locality, where the sponge
passes the remainder of its existence. During this time
some species may never exceed the size of a mustard-seed,
while others attain a diameter of three feet, or even more.
Sponges also vary exceedingly in shape, some having the
form of thin encrusting sheets, others being globular, tubular, cuplike, or highly branched (Fig. 15).

35. **The influence of their surroundings.**—In by far the
larger number of cases an animal possesses the bodily form
of the parent. External agencies may modify this to some
extent, but usually only to a limited degree. A squirrel,
for example, resembling its parent, may grow to a relatively
large or stunted size according to the food supply, and it
may become strong or weak according to the amount of
exercise, and various other changes may result owing to
outside causes; but as a result of these influences the
animal is rarely so modified that one is unable to distinguish
the species. Many of the sponges, however, are exceptions
to this general rule. If, for example, some of the young
of a certain parent develop in quiet water or in an unfavorable locality, they will usually be low, flat, and unbranched; while the others, growing in swiftly running
waterways, develop into tall, comparatively delicate and
highly branched individuals. Under such circumstances
not only does the external form become modified, but
the internal organization may undergo profound change.
The entire organism is plastic and readily molded by
the influence of its surroundings, and the consequent
lack of definite characters often renders it impossible

to assign such forms to a definite position among the sponges.

36. **Structure of a simple sponge.**—In the simpler sponges the body is usually vase-shaped (Fig. 16), with the base fastened to some foreign object, while at an opposite end an opening leads into a comparatively large internal cavity. This latter space is also put in communication with the exterior by a multitude of minute pores which penetrate the body wall. In

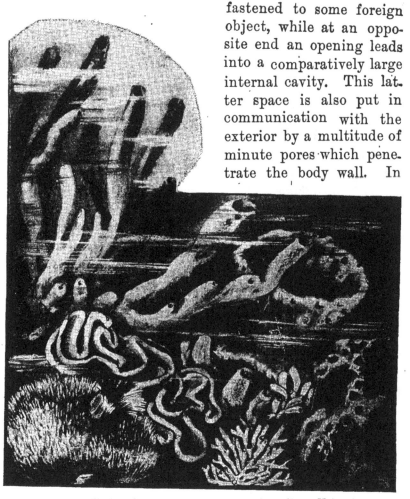

Fig. 15.—Various forms of sponges, natural size. (From Nature.)

the living condition - currents of water continually pass through these smaller canals, and out of the large terminal opening, thus bringing within reach of the body minute floating organisms or organic remains which serve as food. The mechanism by which this process is effected, and the various other structures of the body, are in large part invisible from the exterior, requiring the study of thin sections of the sponge to make them clearly understood.

Under the microscope such a section shows the body of a sponge to consist of an immense number of variously formed cells constituting three distinct layers (Fig. 17). Not only do these layers consist of different kinds of cells, but the duties performed by each are different. For example, a glance at Fig. 17 will show that in the inner layer certain columnar cells exist, provided with a flagellum and encircling collar, the appearance being strikingly like certain of the Protozoa (Fig. 8, B). During life their whip-like processes, lashing backward and forward in perfect unison, produce currents of water which continually pass through the body. The food thus entering the animal is taken up by the cells of the inner layer as it passes by. The supply, however, is usually more than sufficient to meet the demands of this layer, and the excess is passed on to the middle and outer layers. The

FIG. 16.—One of the simplest sponges (*Calcolynthus primigenius* (after HAECKEL). A portion of the wall has been removed to show the inside.

exact method by which this occurs is still a matter of doubt, but there seems to be little question but that each cell of the body receives its food in a practically unmodified condition, requiring that it digest as well as assimilate. The oxygen necessary to this latter process is absorbed by all parts of the body in contact with the water.

37. Skeleton of sponges.—When it is remembered that the protoplasm composing the cells of the sponge has about

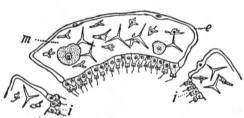

the same consistence as the white of egg, it will be readily understood why the greater number of sponges possess a skeleton. Without such a support the larger globular or branched forms could not exist, and even in the

FIG. 17.—Portion of wall of sponge, showing three layers. *e*, outer layer ; *i*, inner layer, consisting of collared cells ; *m*, middle layer, consisting of irregular cells, among which are the radiate spicules and egg-cells.

smaller members there would be danger of a collapse of the body walls and consequent stoppage of the food supply, owing to the closure of the pores. So in all but a very few thin or flat forms a skeleton appears in the young sponge

almost before growth has fairly begun, and this increases with the body in size and complexity. It is formed by the activity of the cells of the middle layer, and may be composed either of a lime compound resembling marble, or of flint, or of a

FIG. 18.—Different types of sponge spicules.

horn-like substance resembling silk, or these may exist in combination in certain species. When consisting of either of the first-named substances it is never formed in one continuous piece, but of a vast multitude of variously shaped crystal-like bodies termed spicules (Fig. 18). These occur everywhere throughout the body, firmly bound together by means of cells, or so interlocked that they form a rigid support to which the fleshy substance is bound and through which the numerous canals penetrate.

In a relatively few species only does the skeleton consist of horn, though there are many in which horn and flint exist together. In the former event, if the skeleton be elastic and of sufficient size, it becomes valuable to others than the naturalist, for the familiar sponges of commerce are the horny skeletons of forms usually taken in the West Indies or in the Mediterranean Sea. In these localities the animals are pulled off by divers, or with hooks, and are then spread out in shallow water where the protoplasmic substance rapidly decays. The remaining skeleton, thoroughly washed and dried, is ready for the markets of the civilized world.

Examining a bit of such a " sponge " under a magnifying glass, it will be seen that the skeleton is not composed of various pieces, but of one continuous mass of branching fibers, which interlace and unite in apparently the greatest confusion; yet in the living animal these were perfectly adapted to the position of the canals and the general needs of the animal.

Besides being a scaffold-work to which the fleshy portions of the body are fastened, the skeleton serves also for protection. In some species, needle-like spicules as fast as they are formed are partly pushed out over the entire surface of the body, giving the appearance of a spiny cactus; or in other cases they are arranged in tufts about the canals, effectually preventing the entrance of any marauder. Thus perfectly protected, the sponges have but few natural enemies, and hence it is that in favorable localities they grow in great profusion.

38. Race histories and life histories.—We have now traced living things from their simplest beginnings, where they exist as single cells, and have seen that in bygone times similar forms have united into simple colonies, and these

through a division of labor among the constituent cells have resulted in *Volvox*-like colonies. There are the strongest reasons for the belief that as these simple forms scattered into various surroundings and underwent changes to meet the shifting conditions, they assumed different degrees of complexity that have resulted in the animal forms of the present day.

It may have been noticed also that the sponge in its development passes through these stages : a single-celled egg ; later, a young form similar to *Pandorina*, then growing to look like *Volvox*, and finally assuming its permanent form. The history of the race of sponges and their development through a long line of ancestry of increasing complexity is thus told by the sponge as it develops from the egg into the adult; and, so far as we know, all the many-celled animals in their growth from the egg repeat more or less clearly the stages passed through by their forefathers.

CHAPTER V

THE CŒLENTERATES

39. **General remarks.**—This division of the many-celled animals includes the jelly-fishes, sea-anemones, and corals. A few species live in fresh water, but the majority are confined to the sea, being found everywhere from the shore-line and ocean surface to the most profound depths. Adapted to different surroundings and modes of life, they constitute a vast assemblage of the most bewildering diversity. In some cases their resemblance to plants is remarkable, and the term zoophyte or "plant animal," occasionally applied to them, is the relic of former times when naturalists confounded them with plants. Even to-day certain species are sometimes collected and preserved as seaweeds by the uninformed.

The general plan on which all cœlenterates are constructed is a simple sac, in some respects resembling that of the lower sponges, yet, since the modes of life of the members of the two groups are usually quite unlike, we shall find many profound differences between them.

40. **The fresh-water Hydra.**—The bodily plan comes out most clearly in the *Hydra* (Fig. 19, A, D), which occurs upon the stems and leaves of submerged fresh-water plants in this and other countries. Its body, of a green or grayish color, according to the species, scarcely ever attains a diameter greater than that of an ordinary pin nor a length exceeding half an inch. One end of the cylindrical organism is attached to some foreign object by means of a sticky secretion, but as occasion requires it may free itself, and by

means of a "measuring-worm" movement travel to another place.

Examined under a hand lens, the free end of the body will be found to support six to eight prolongations known

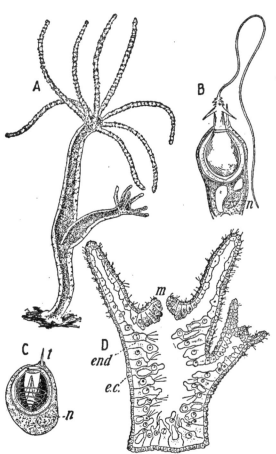

as *tentacles*, which serve to convey food to the mouth, centrally located in their midst. This opening, unlike that of the sponges, is the only one leading directly into the large central gastric cavity which occupies nearly the entire animal (Fig. 19, D). As in the sponge, the cells of the body are arranged in the form of definite layers, but the middle one is represented only by a thin gelatinous sheet.

41. **Organs of defense.** — These are the so-called *lasso* or *nettle-cells* (Fig. 19, C). Some

FIG. 19.—The fresh-water *Hydra*. A, entire animal, developing a new individual (enlarged 25 times). B, C, nettle-cells (after Schneider) ; D, section through the body.

of the cells of the outer layer possess, in addition to the elements of the typical cell, a relatively large ovoid sac filled with a fluid, and also a spirally wound hollow thread provided with barbs near its base. On the outer extremity of the nettle-cell projects a delicate bristle-like process, the *trigger hair*. These cells are especially abundant on the tentacles (Fig. 19, A, D), forming close, knob like elevations or "batteries," thus rendering it practically impossible for any free-swimming organism to avoid touching them in brushing against the tentacles. In such an event the disturbances conveyed through the trigger hair set up in some unknown way very rapid changes in the cell. This causes the sac to discharge the coiled thread and barbs into the body of the intruder, which is rendered helpless by the paralyzing action of the fluid conveyed through the thread. Thus benumbed it is rapidly borne to the mouth and swallowed. In time new nettle-cells develop to take the place of those discharged and consequently worthless.

42. **Digestion of food.**—Upon the interior of the body of *Hydra* and all of the cœlenterates the food, by reason of its large size, is incapable of being taken into the various cells. It is necessary, therefore, to break it up into smaller masses, and this is accomplished through the solvent action of the digestive fluid poured over it from some of the cells of the adjacent inner layer. When subdivided, the granules swept about the gastric cavity by the beating of the flagella (Fig. 19, D) are seized by the processes on the free surfaces of the remaining inner layer cells, where they undergo the final stages of digestion; then in a dissolved state they become absorbed and assimilated by all the cells of the body.

43. **Methods of multiplication.**—Very frequently, especially if the *Hydra* has been well fed, two or three processes arising as outpushings of the body wall may be noted upon the sides of the animal (Fig. 19, A, D). If these be watched from time to time they are found to increase in size, and finally, upon their free extremities, to develop a mouth and surrounding tentacles. Up to this point growth has taken place as a result of the assimilation of nutritive substances supplied from the parent; but a con-

striction soon occurs which separates the young from the parent, and from that time on the two lead independent existences. At other times this *asexual* method of multiplication is replaced by *sexual* reproduction, where new individuals arise from fertilized eggs. Both eggs and sperm arise in *Hydra* and in some other animals in the same individual, but in all such cases the eggs are fertilized by sperm which escape from some other individual. The fertilized egg, surrounded by a firm coat, separates from the parent, drops to the bottom, and after a period of rest develops into a little *Hydra* which hatches and enters upon a free existence.

FIG. 20.—Different types of Hydrozoan colonies. From Nature. the lower species magnified about 50 diameters.

44. Hydrozoa, or Hydra-like animals.—Attention has already been directed to the fact that the structure of *Hydra* is the simplest of the cœlenterates; nevertheless, the thousand or more species belonging to this class which present a much more complicated appearance (Fig. 20) possess many fundamental *Hydra*-like characters. It is owing to this fact that this assemblage of forms has been placed in the class of the Hydrozoa, or *Hydra*-like animals.

With but very few exceptions the members of this class are marine, usually living near the shore-line, where at times their plant-like bodies occur in the greatest profusion attached to rocks, seaweeds, or the bodies of other animals, particularly snails and crabs. Fig. 20 (upper colony) gives a good idea of one of the more complex forms, whose tree-like body attains in some cases the relatively giant height of from 15 to 25 c.m. (six to ten inches). In early life it bears a close resemblance to a *Hydra*. Buds form in much the same way, but they retain permanently their connection with the parent, and in turn bear other buds, until finally the form shown in the figure is attained. In the meantime root-like processes have been forming which afford firm attachment to the object upon which the body rests. Also during this process the cells of the outer layer form a horny external skeleton ensheathing the entire organism except the terminal portions (the hydranths, Fig. 21, B) bearing the tentacles. The gastric cavities of all communicate, and the food captured by one ministers in part to its own needs and, swept through the tubular stalks and roots, is also shared by all other members.

45. Jelly-fishes and the part they play.—During the process of growth a number of stubby branches arise which differ from the ordinary type in shape, and also in many cases as regards color. These club-like, fleshy portions develop close-set buds (Fig. 21, c) which early assume a bell-like shape, the point of attachment corresponding to the handle, while the clapper is represented by a short, slender process bearing on its end an opening which becomes the mouth (Fig. 21, A). Around the margin of the bell numerons tentacles develop, and at the same time the gelatinous substance situated between the outer and inner layers of the bell expands to a relatively enormous degree, giving it an increasing globular form and glassy appearance.

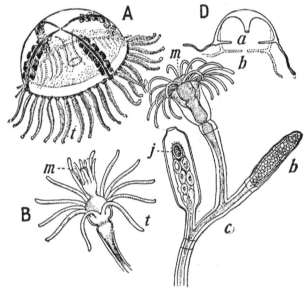

Fig. 21.—A jelly-fish (*Gonionemus*), slightly enlarged. The stalked mouth is shown in dotted outline. B, C, enlarged portions of a hydroid colony bearing the mouth and tentacles; *j*, a capsule within which the jelly-fish develop; D, diagram of jelly-fish, illustrating its method of locomotion.

Finally, vigorous movements rupture the connection with the parent, and this newly developed outgrowth, usually small, becomes an independent organism popularly termed a jelly-fish. While the external form of the jelly-fish appears to be widely different from the hydranths, a more careful study shows the difference to be superficial. Some zoologists believe that jelly-fishes are simply buds which have become fitted to separate and swim away from the colony in order to distribute the young, as described hereafter.

When the stalked colonies are very abundant the jelly-fishes may be liberated in such multitudes that the upper surface of the ocean for many miles may be closely packed with them in numbers reaching far into the millions. In these positions they are carried both by oceanic currents and through the alternate expansion and contraction of the bell, a movement resembling the partial closing and opening of an umbrella. In the jelly-fish the contraction is more vigorous and rapid than the opening in the velum or veil (Fig. 21, *b*) which is so narrowed that the water in the subumbrella space (*a*) is driven through it with considerable force, which results in driving the body in the opposite direction.

The life of a jelly-fish is perhaps of short duration, lasting not more than a few hours in some species, up to two or three weeks in others, but during that period they produce multitudes of eggs which develop into minute free-swimming young. These settle down on some rock or seaweed, and soon develop a *Hydra*-like body which, after the fashion described above, grows into another tree-like colony.

46. **Alternation of generations.**—It will be noticed that the offspring of the jelly-fishes are not jelly-fishes, but stalked colonies, and these latter forms give rise to jelly-fishes. This is known as the *alternation of generations*, the jelly-fish generation alternating with the colonial form. This characteristic is of the greatest service in preventing the extermination of the race. Were the stalked forms to give rise directly to other stationary colonies, it is obvious that before long all the available space in the immediate locality would be filled. The food supply, always limited, would not suffice, and starvation of some or imperfect development of all would result; but by means of the free-swimming jelly-fish new colonies are established over very extensive areas, and favorable situations are held by all.

47. **More complex types.**—As mentioned above, there are perhaps upward of a thousand species of Hydrozoa, all with

essentially the same structure but with various modes of branching (for some of the commoner modes, see Fig. 20). In some of the higher forms a division of labor has arisen among various members of the association which has led to most interesting results. For example, Fig. 22 represents a species of hydroid found investing the shells of sea-snails occupied by hermit crabs (Fig. 66). To the unaided eye its appearance is that of a delicate vegetable growth, but when placed under the microscope it is found to consist of a multitude of *Hydra*-like animals united by a hollow branching root system connecting the gastric cavities of all of them (Fig. 22). Certain individuals (*a*) with tentacles and a mouth resemble a *Hydra*; others, without a mouth and tentacles, are reduced to a club-like form (*b*) liberally supplied with nettle-cells upon their free extremities; while the third type

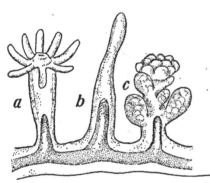

FIG. 22.—An enlarged portion of a hydroid colony (*Hydractinia*), showing (*a*) the nutritive polyp, (*b*) the defensive polyp, and (*c*) the reproductive polyp.

(*c*), likewise devoid of a mouth, possesses rudiments of tentacles below which are borne numerous clumps of reproductive cells. The first type, the only one possessing a mouth, captures the food, and after digesting it distributes the greater portion to the remaining members by means of the connecting root system; those of the second form, defending the others by means of their nettle-cells against the inroads of a foreign enemy, are the soldiers of the colony; while the third type produces the eggs from which new individuals develop.

In some of the higher Hydrozoa, the Portuguese man-of-war (Fig. 23), this division of labor has reached a more advanced stage of development, and in addition the entire colony is fitted for a free-swimming existence. What corresponds ordinarily to the attached stalk in other forms terminates in a bladder-like expansion, distended with gas, that serves as a float. From it are suspended individuals resembling great streamers sometimes many feet in length, without mouths, but loaded with nettle-cells that enable them to capture the food, which is conveyed to the second type, the nutritive polyps. Each of these is a simple tube bearing a mouth, and within them the food is digested and distributed by means of a branching gastric cavity extending throughout the entire colony. Then there are individuals like mouthless jelly-fishes which bear the eggs and care for the perpetnation of the colony; and besides these there may be some whose duty it is to defend the rest, and others whose active swimming movements, together with. the wind, drive the colony about. Thus united, sharing the food supply and working for the general welfare of all, the members of this colony live in greater security and with less effort than if, as separate individuals, each was fighting the battles of life alone.

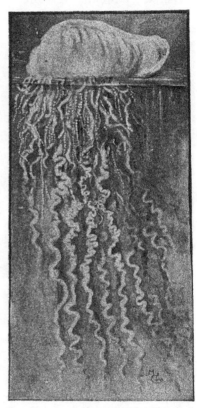

FIG. 23.—A colonial jelly-fish (*Physalia*). From Nature.

48. **Scyphozoa.**—The greater number of the larger and more conspicuous jelly-fishes are included under this term. In general shape and locomotion they resemble those of the

preceding group (Fig. 24), but while the latter are generally very small, these forms are commonly from four to twelve inches in diameter, and some measure one to two meters (three to six feet) across the bell. They are also distinguished by means of tentacles which extend from the corners of the mouth sometimes to a distance of several feet, and together with the marginal tentacles are formidable weapons for capturing small crabs, fishes, and other animals which serve as food. In turn these forms serve as the food of many whales, porpoises, and numerous fishes which hunt them down, though the amount of nourishment they contain is probably relatively small owing to the fact that in their composition there is a large percentage of water (99 per cent in some species).

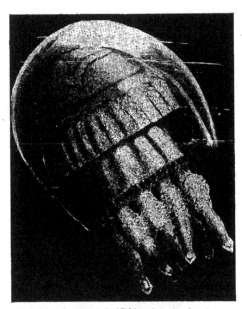

FIG. 24.—A jelly-fish (*Rhizostoma*), about one-fourth natural size.

The lobed margin of the bell, the absence of a definite swimming organ or velum, and the character of several of the internal organs, distinguish the larger from the smaller jelly-fish ; but the greatest difference, however, is in the method of development.

49. **Development.**—The eggs arise from the inner layer of the jelly-fish and drop into the gastric cavity, where each develops into a ciliated two-layered sac in some respects like that of a young sponge. Swimming away from the parent, they finally settle down, and attaching themselves (Fig. 25, *a*) assume the external form and habits of the sea-

anemones, described in the next section. In the course of time remarkable changes ensue, which first manifest them-

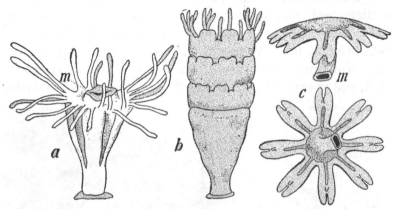

FIG. 25.—Stages in the development of a scyphozoan jelly-fish. *a*, the attached young, which in *b* has separated into a number of disks, each of which becomes a jelly-fish, *c*.—After KORSCHELT and HEIDER.

selves in a series of grooves encircling the body. These grow deeper, and the body of the animal finally comes to resemble a pile of saucers with the edge of each developed into a number of lobes (Fig. 25, *b*). One after another each saucer, to preserve the simile, raises itself from the top of the pile and swims away, and is clearly seen to be a jelly-fish, though considerably unlike the adult. As growth pro-ceeds, however, it un-dergoes a series of transformations which result in the adult form.

FIG. 26.—An attached scyphozoan jelly-fish (*Haliclystus*). Natural size, from Nature.

50. Sea-anemones.—In its external appearance the sea-anemone (Fig. 27) bears some resemblance to the *Hydra*, but is of a much larger size (1 to 45 c.m., or ½ inch to 1½ feet in diameter), and is frequently brilliantly colored. The number of tentacles is also more numerous, and the mouth leads into the body by means of a slender esophagus (Fig. 28). Numerous partitions from the body wall extend inward, and many unite to the esophagus, keeping the latter

Fig. 27.—Sea anemones (the two upper figures) and solitary coral polyps.

in position. Below the esophagus each partition projects into the great cavity of the body and bears upon its inner free edge several important structures. The first of these, known as the mesenteric filaments (Fig. 28), appearing like delicate frills, plays an active part in the digestion of the food. Associated with these are long, slender threads, closely packed with innumerable lasso-cells, which may be thrown out through openings in the body wall when the animal is attacked. Lasso-cells are also very numerous on the tentacles, which are thus to some extent defensive, but are chiefly active in capturing the crabs and small fish which serve as food.

The partitions also carry eggs which may undergo the first stages of their growth within the body, and when finally able to swim are sent out through the mouth opening by hundreds to seek out favorable situations, there to settle down and remain. In some species the young may sometimes arise as buds, as in *Hydra* (Fig. 27), and in others the animals have been described as splitting longitudinally into two equal-sized young.

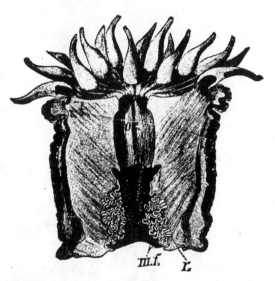

FIG. 28.—Longitudinal section through the body of a sea-anemone. *oe*, esophagus ; *m. f.*, mesenterial filaments ; *r.*, reproductive organs.

51. Corals.—The coral polyps also belong to this group, showing a very close resemblance to the sea-anemones. In most cases they develop a firm skeleton of lime, commonly known as "coral," which serves to protect and support the body. In a few species the polyps throughout life are solitary, and with skeleton comparatively simple (Fig. 27) ; but the larger number of species become more complex by developing buds, which retain their connection with the parent, and in turn produce other outgrowths with the ultimate result that highly branched colonies are produced (Fig. 29). At the same time the outer layer of the body is continually forming a skeleton which encloses the colony as a sheath, except at the termination of each branch, where the mouth and tentacles are located. In certain species—for example, the sea pens (*Pennatula*) and sea fans (*Gorgonia*)—a skeleton may be

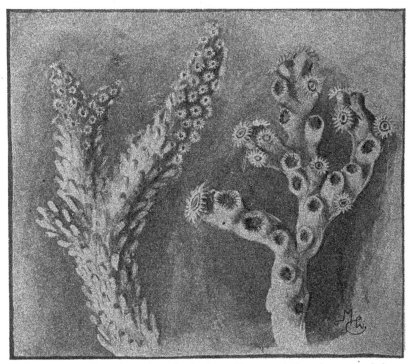

Fig. 29.—Small portions of coral colonies, with some of the polyps expanded.

formed of myriads of lime spicules, somewhat like those of the sponge, which are bound together by the fleshy substance of the body; but the skeleton of most of the common forms in the ocean, and the coral found· in general collections, is stony. According to their method of branching, such specimens have received various popular names, such as brain, stag-horn, organ-pipe, and fungous corals.

Fig. 30.—Coral island (Nanuku Levu, of the Fiji group). (After a photograph by MAX AGASSIZ.)

Fig. 31.—Shore of a coral island, with cocoanut palms. (After a photograph.)

Nearly all species, like the sea-anemones, are brilliantly colored during life, and several are highly phosphorescent. All are marine, and while they are found everywhere, from the shore-line to great depths, the more abundant and larger species inhabit the clear, warm waters of the tropics down to a depth of one hundred and sixty feet. In such regions the stag-horn corals especially grow in the wildest profusion, and become tall and greatly branched. Except in quiet water they are continually being broken by the waves, beaten into fragments, and the resulting sand is deposited about their bases. As a result of this continuous growth and erosion, there have been formed from coral sand mixed with the shells of mollusks and the skeletons of various Protozoa several of the islands along the Florida coast and many of those of the Pacific, some of them hundreds of miles in extent.

CHAPTER VI

THE WORMS

52. General Characteristics.—The bodies of the animals comprising the two preceding groups are exposed on all sides equally to the water in which they live and are radially symmetrical; but in the worms, one side of the body is fitted for creeping, and for the first time we note a well-marked dorsal (back) and ventral (under) surface. In the former, the body, like a cylinder, may be divided into similar halves by any number of planes passing lengthwise through the middle; but in the worms, the right and left halves only are exposed equally to their surroundings, and there is, accordingly, only one plane which divides the body into corresponding halves, so that these animals, like all higher forms, are bilaterally symmetrical. In creeping, also, one·end of the body is directed forward and it thus becomes correspondingly modified. It usually bears the mouth, and may be provided with eyes, feelers, or organs of touch, and various other structures which enable the worm to recognize the nature of its surroundings. The nervous and muscular systems are better developed than in the foregoing groups, and we note a greater vigor and definiteness in the animal's movements, and in various ways the worms appear better able to avoid or ward off their enemies, recognize and select their food, and in general adapt themselves to the conditions of life.

The division of the worms is a very large one, and in some respects difficult to define, owing to the close resemblance which many of them show to animals in other groups. All the invertebrates, therefore, except the crabs and insects, were placed in one group until subsequent study made it possible to classify them more exactly. According to the general shape of the body, and the arrangement of internal organs, worms are divided into a number of groups, chief among which are the flatworms, the thread or roundworms, and the ringed worms or annelids.

THE FLATWORMS

53. Form and habitat.—The flatworms, as their name indicates, are much flattened, leaf-like forms, some species living in damp places on land, in fresh - water streams or ponds, or along the seacoast, while a variety of other species are parasitic. The free forms (Fig. 32) are usually small, barely reaching a length greater than five or seven centimeters (2 to 3 inches), but some of the parasitic species (Fig. 36) attain the great length of six to thirteen meters (20 to 40 feet).

The free-living forms usually occur on the under side of stones, and frequently are so delicate that a touch is sufficient to destroy them. A few species are almost transparent, while many are colored to harmonize completely

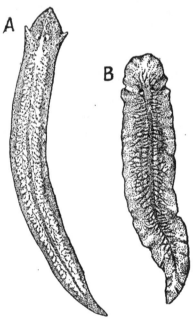

FIG. 32.—A, fresh-water flatworm (*Planaria*); B, marine flatworm (*Leptoplana*). Enlarged, from Nature.

with their surroundings, so that, even though fragile and defenseless, they escape the attacks of enemies by being overlooked. The night-time or dark days are their hunting season, and at such periods they may be found moving about with a steady gliding motion (due to cilia covering the entire body), varied occasionally by a looping, caterpillar movement, or by swimming with a flapping of the sides of the body. When watched at such times they may sometimes be seen to snatch up small worms, snails, small crabs and insects, which serve as food.

More closely examining one of these forms, for example, the species usually found on the under side of sticks and stones in our shallow fresh-water streams (Fig. 32, A), we note that the forward end is not developed into a well-defined head as in the higher worms, but is readily determined by the presence of very simple eyes and tentacles, while the lower creeping surface is distinguished by a lighter color and the presence of the mouth. Through this small opening a slender proboscis (in reality the pharynx) may be extended some distance, and may be seen to hold the small organisms upon which it lives until they are sufficiently digested to be taken into the body.

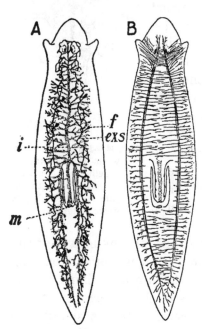

Fig. 33.—Anatomy of fresh-water flat-worm (*Planaria*). *exs*, excretory system, with flame-cell (*f*). The alimentary canal is stippled. B, nervous system.

54. **Digestive system.**—In the smaller flatworms, some of which are scarcely larger than many of the Protozoa, the alimentary canal is a simple unbranched tube; but in the larger forms such an apparatus is replaced by a greatly branched digestive tract which furnishes an extensive surface for the rapid absorp-

tion of food, and extending deep into the tissues of the body, carries nutriment to otherwise isolated regions. In the fresh-water forms and their allies there are three main branches of the intestine (Fig. 33), while in many of those from the sea there are several, and their arrangement affords a basis for their general classification.

55. **Excretory system.**—In the sponges and cœlenterates the wastes are cast out by the various cells into the gastric cavity or at once to the exterior without the aid of any pronounced system of vessels; but in the flatworms several of the organs are deeply buried within the tissues of the body and a drainage system becomes a necessity. This consists of a paired system of vessels extending the length of the animal (Fig. 33) and provided with numerous branches, some of which open at various points on the surface of the body, while the others terminate in spaces (Fig. 34, s) among the organs in what are known as flame-cells. The substances which accumulate in these spaces are gathered up by the flame-cell, poured into the space it contains, and by means of the vibratory motion of its flagellum (f), a movement bearing a fancied resemblance to the flickering of a flame in the wind, are borne through the tubes to the exterior.

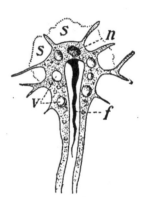

FIG. 34.—Flame-cell of flatworm (after LANG). f, flagellum; n, nucleus; s, spaces among the organs of the body; v, waste materials.

56. **Nervous system and sense-organs.**—In the sponges no definite nervous system is known to exist, the slight movements which the cells are able to undergo being regulated somewhat as they are in the Protozoa. Among the cœlenterates certain of the cells scattered over the surface of the body are set aside as nerve-cells, and, more or less united by means of fibers extending from them, convey impulses over the body. In the flatworms the larger number of nerve-cells

are collected into two definite masses (Fig. 33, B), which constitute a simple brain on which the eyes are situated and from which bundles of nerve fibers pass to all parts of the body, the two extending backward being especially noticeable. As in the squirrel, these are distributed to the muscles and other organs to regulate their activity, while those distributed to the skin, especially in the forward part of the body, convey stimuli produced by touch. The branches connecting with the eyes enable the animal to distinguish light from darkness, but are probably too simple to allow it to clearly distinguish objects of the outside world. The sense of smell and possibly that of taste are also present, but are relatively feeble.

Some other characters of this class will be noted in the consideration of the two following classes.

57. **Parasitic flatworms (trematodes)—parasitism.**—Mention has already been made of the associations of two animals as "messmates" for mutual benefit, such as the Hydractinia growing on the surface of the shell inhabited by the hermit-crab, to which it gives protection by means of its nettle-cells, while in turn being borne continually into regions abounding with food. More frequently, however, one animal derives benefit from another without making any compensation. For example, many species of flatworms live within the shells of certain snails and upon the bodies of sea-urchins and starfishes, where they gather in their food supply safe from the attacks of enemies. Such associations are probably without much if any inconvenience to the animal thus inhabited, and it also appears probable that the tenants are transients, using the mollusk or starfish only as a temporary home. But from this condition of affairs it is only a short step to the parasitic habit, where the association becomes permanent and the occupant is provided with various structures which prevent its separation from its host. This latter kind of union occurs throughout the group of trematodes; all are parasitic, and

their internal organization, so closely resembling that of the free-living forms as to need no further description, indicates that they are descendants of the latter. In the greater number the body is flat, and a few species still retain their outer coat of cilia; but since these are no longer of service as locomotor organs they have generally disappeared, and in their place• numerous adhesive organs, such as spines, hooks, and suckers (Fig. 35), have arisen, which enable the animals to hold on with great tenacity. Thus attached to its host, and using it as a convenient and comparatively safe means of locomotion, the parasite may still continue to capture small animals for food or may derive its nourishment from the tissues of the host. In addition there are numbers of internal parasites, living almost exclusively in the bodies of vertebrate animals, scarcely a single one escaping their ravages.

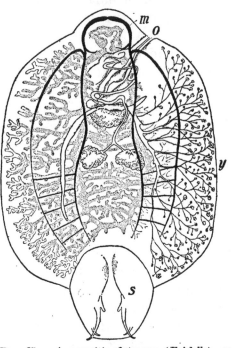

Fig. 35. — A parasitic flatworm (*Epidella*). *m* mouth; *o*, opening of reproductive system; *s*, sucker and spines for attachment. The digestive system is stippled; nervous system black. Enlarged 8 times, from Nature.

58. **Life history.**—In the external parasites the young hatch out and with comparative ease make their way to another host; but the young of an internal parasite, inhabiting the alimentary canal, have a very slight chance indeed of ever reaching a similar location in another host.

For this reason an almost incredible number of eggs is laid, and some extraordinary measures are employed in effecting the desired result. Probably the best-known example is that of the liver fluke inhabiting the bile-ducts in the sheep. Each worm lays several hundred thousand eggs, which make their way from the host, and if they chance to fall in pools of water or damp situations may proceed to develop, otherwise not. If the surroundings be favorable, the young, like little ciliated Infusoria, escape from their shells and restlessly swim or move about for a short time, and if during this time they come in contact with certain species of snails living in these situations they at once bore into their bodies. Here they produce other young somewhat resembling a tadpole, that now make their escape from the snail. In a short time each one crawls upon a blade of grass, and surrounds itself with a tough shell, where it may remain for several weeks. If the grass on which they rest be eaten by a sheep, they finally make their way to the bile-ducts and there become adult. The life cycle is now complete; the young form has found a new host; and the process shows how wonderfully animals are adapted to the conditions which surround them, and how closely they must conform to these conditions in order to exist.

59. **The tapeworms (cestodes).**—The cestodes, or tapeworms, are also parasitic flatworms in which the effects of such a mode of life are strongly marked. They occur almost exclusively in the bodies of vertebrate hosts and exhibit a great variety of bodily forms, in some cases resembling rather closely the trematodes, but in others strikingly different. In the latter type the body is usually of great length (from a few centimeters to upwards of sixteen meters (50 feet)), and terminates in a "head" (Fig. 36) provided, in the different species, with a great variety of hooks and spines and numbers of suckers for its attachment to the body of the host. From the head the body extends backward in the gradually enlarging ribbon-like body, slender at

first and scarcely showing the segments which finally become so prominent a feature.

When carefully examined, a two-lobed brain is found
in the " head," and from it nerves extend the entire length
of the body, followed throughout their
course by the tubes of the excretory
system ; also each segment contains a
perfect reproductive system, so that
even if it be separated from the others
it may continue to exist for a considerable length of time. Furthermore,
the tapeworms are surrounded by the
predigested fluids of their host; a
special alimentary canal is therefore
superfluous, and all traces of it have
disappeared.

60. **Development.**—As the animal
clings in this passive way to the body
of its host the segments, loaded with
eggs ready for development, separate
one after another from the free end
of the body, pass to the exterior, and

Fig. 36.—Tapeworm (*Tænia
solium*). In upper left-
hand corner of figure is
the much enlarged head.
—After Leuckart.

slowly crawling about like independent organisms, lay great
numbers of eggs, which may find an intermediate host as in
the life cycle of the liver fluke, and so in time find their
permanent resting-place. Fortunately in all these parasitic
forms, though an inconceivably great number of eggs are
laid, only a comparatively few reach maturity. Even these,
however, may cause at times great destruction among the
higher, and especially our domestic, animals, often doing
damage amounting to many millions of dollars per year.

61. **The tapeworm in relation to regeneration.**—It has
been known for more than one hundred and fifty years that
some of the lower animals possess to a surprising degree
the ability to regenerate parts of the body lost through
injury. The *Hydra*, hydroids, and some of the jelly-fishes

may be cut into a number of pieces, each of which will develop into a complete individual; and this power of recovery from the injuries produced by enemies is of the greatest service in the perpetuation of the species. This ability is also present in certain flatworms, and some species are known which voluntarily separate the body into two portions, each of which becomes an adult. In other species a similar process results in the formation of a chain of six individuals, placed end to end, the chain finally breaking up into as many complete worms. It is possible that the tapeworm may also be looked upon as a great chain of united individuals produced by the division of a single original parent, which becomes adapted for attaching the others until they separate. These latter are capable only of a very sluggish movement, and, devoid of mouth and alimentary canal, are not able to digest their food, but their life work is to so lay their eggs that they may develop into other individuals, and for this they are well adapted.

Nematodes (Threadworms)

62. General characters.—This class of worms is composed of an enormous number of different species, some parasitic, others free all or a portion of their lives, and in view of the fact that they inhabit the most diverse situations it is remarkable that they are so uniform in their structure. In all the body is slender, and the general features of its organization may be readily understood from an examination of the " vinegar eel " (Fig. 37, A). This small worm (not an eel), a millimeter or two in length, lives on the various forms of mold that grow in fermenting fruit juices, especially after a little sugar or paste has been added. A tough cuticle surrounds the body, preserving its shape and at the same time protecting the delicate organs against the action of the acids in which it lives. Through this may be seen great bands of muscles extending the entire length of the body and producing the wriggling movements of swimming or crawling. They also give support to a brain, which is in the form of a collar encircling the pharynx near the head, and to the great nerves which extend from it. Still further within the transparent body the alimentary canal may be distinguished as a straight tube passing directly through the animal. The alimentary canal lies freely in a great space, the body cavity, traces of which may exist in the flatworms in the form of hollow spaces into which the kidneys open. It is possible that in this form also the kidneys open into this space, and it is roomy enough besides to afford lodgment for the reproductive organs in addition to a large amount of fluid which is probably somewhat of the nature of blood. A space in some respects similar to this occurs in all the animals above this group, and as we shall see, it is often curiously modified and serves for a number of different and highly important purposes. In the roundworms the fluid it contains probably acts in the nature of a blood system, distributing the food and

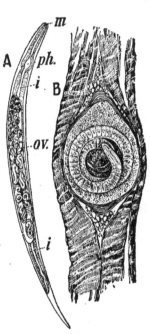

FIG. 37. — Thread- or round-worms. A, vinegar eel (*Anguillula*) ; *m*, mouth ; *ph.*, pharynx ; *i*, intestine ; *ov.*, developing young. B, *Trichina*. From Nature, greatly enlarged.

oxygen to various parts of the body and carrying the wastes to the kidneys for removal.

63. **Multiplication.**—In the matter of the production of new individuals the greatest differences exist. In some threadworms, for example the "vinegar eel," eggs develop within the body and the young are born with the form of the parent. In other cases the eggs are laid in the water, where they, too, may directly grow to the adult condition ; but in

the greater number of species the development is round-about, and one or more hosts are inhabited before the young assume the adult condition. Such is the case with the dreaded *Trichina* (Fig. 37, B), which infests the bodies of several animals, particularly the rat. When these forms are introduced into the alimentary canal of the rat, for example, they soon lay a vast quantity of eggs, sometimes many millions, which develop into young that bore their way into the muscles of the body, where they may remain coiled up for years. If the body of the rat be eaten by some carnivorous animal, these excessively small young are liberated during the process of digestion and rapidly assume the adult condition in the alimentary canal, likewise giving rise to young which pursue again the same course of development.

Another example of a complicated life history is in the *Gordius* or "horsehair snake" (a true worm and not a snake) frequently seen in the spring in pools where it lays its eggs. These eggs develop into young which bore their way into different insect larvæ, which are in turn eaten by some spider or beetle, and the worm thus transferred to a new host. In this they grow to a considerable size, and then make their exit from the body of the host and finally become adult.

64. **Spontaneous generation.**—The ancients believed that many animals were spontaneously generated. The early naturalists thought that flies arose by spontaneous generation from the decaying matter of dead animals; from a dead horse come myriads of maggots which change into flesh flies. Frogs and many insects were thought to be generated spontaneously from mud. Eels were thought to arise from the slime rubbed from the skin of fishes. Aristotle, the Greek philosopher, who was the greatest of the ancient naturalists, expresses these beliefs in his books. It was not until the middle of the seventeenth century— Aristotle lived three hundred and fifty years before the

birth of Christ—that these beliefs were attacked and began to be given up. William Harvey, an English naturalist, declared that every animal comes from an egg, but that
the egg might "proceed from parents or arise spontaneously or out of putrefaction." In the middle of the same
century Redi proved that the maggots in decaying meat
which produce the flesh flies develop from eggs laid on the
meat by flies of the same kind. Other zoologists of this
time were active in investigating the origin of new individuals. And all their discoveries tended to weaken the
belief in the theory of spontaneous generation.

Finally the adherents of this theory were forced to
restrict their belief in spontaneous generation to the case
of parasites and the animalcules of stagnant water. It was
maintained that parasites arose spontaneously from the
matter of the living animal in which they lay. Many parasites have so complicated and extraordinary a life history
that it was only after long and careful study that the truth
regarding their origin was discovered. No case of spontaneous generation among parasites is known. If some
water in which there are apparently no living organisms,
however minute, be allowed to stand for a few days, it will
come to be swarming with microscopic plants and animals.
Any organic liquid, exposed for a short time, becomes
foul through the presence of innumerable bacteria, etc.
But it has been certainly proved that these organisms are
not spontaneously produced by the water or organic liquid.
A few of them enter the water from the air, in which there
are always greater or less numbers of spores of microscopic
organisms. These spores germinate quickly and the rapid
succession of generations soon gives rise to the hosts of
bacteria and Protozoa which infest all standing water.
If all the active organisms and inactive spores in a glass
of water are killed by boiling the water, "sterilizing" it,
as it is called, and this sterilized water be put into a
sterilized glass, and this glass be so well closed that germs

or spores can not pass from the air without into the sterilized liquid, no living animals will ever appear in it. It is
now known that flesh will not decay or liquids ferment
except through the presence of living animals or plants.

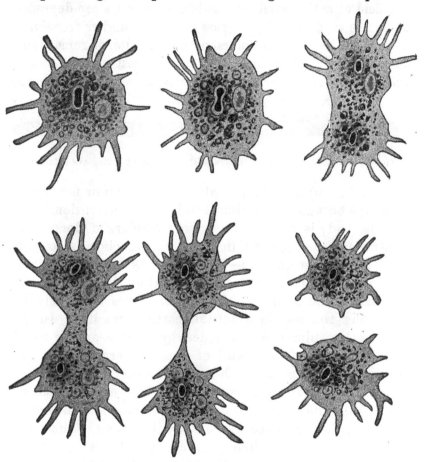

FIG. 38.—The multiplication of *Amœba* by simple fission.

To sum up, we may say that we know of no instance of the
spontaneous generation of organisms, and that all the animals whose life history we know are produced from other
animals of the same kind. " *Omne vivum ex vivo*," All life
from life.

ANNELIDS OR SEGMENTED WORMS

65. The earthworms and their relatives.—Leaving the groups of the parasitic animals, which have been driven from the field of active existence and in many ways are degraded by such a mode of life, we pass on to the higher free-living worms, where brilliant colors, peculiar habits, or remarkable adaptations render them peculiarly interesting. In considering first their general organization, we may use the earth-

FIG. 39.—Earthworm (*Lumbricus terrestris*). *m*, mouth ; *c*, girdle or clitellum.

worm (Fig. 39) (sometimes called angle-worm or fish-worm) as a type because of its almost universal distribution.

The body is cylindrical, shows well-marked dorsal and ventral surfaces, and, as in all of the annelids, is jointed, each joint being known as a *segment*. Anteriorly it tapers to a point, and the head region bearing the mouth is ill-defined, unlike many sea forms, yet serves admirably for tunneling the soil in which all earthworms live. In this process the animal is also aided by bristles or *setæ* which project from the body wall of almost every segment and may be stuck into the earth to afford a foothold.

66. Food and digestive system.—The earthworms are nocturnal animals, seldom coming to the surface during the day except when forced to do so by the filling of their tunnels with water or when pursued by enemies. At night they usually emerge partially, keeping the posterior end of the body within the burrow, and thus they scour the surrounding areas for food, which they appear, in some cases at least, to locate by a feeble sense of smell. They also frequently extend their habitations, and in so doing swallow enormous quantities of earth from which they digest out any nutritive substances, leaving the indigestible matter in coiled "castings" at the entrance of the burrows. In thus mixing the soil and rendering it porous they are of great service to the agriculturist.

Although earthworms are omnivorous they also manifest a preference for certain kinds of food, notably cabbage, celery, and meat, which leads us to think that they have a sense of taste. All these substances are carried into their retreats and devoured, or are used to block the entrance during the day. The food thus carried into the body is digested by a system (Fig. 40) composed of several portions,

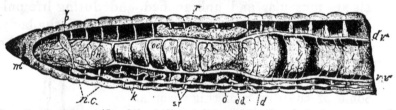

Fig. 40.—Earthworm (*Lumbricus*) dissected from left side. *b*, brain ; *c*, crop ; *d*, outer opening of male reproductive system ; *dv*, dorsal blood-vessel ; *g*, gizzard ; *h*, pulsating vessels or "hearts" ; *i*, intestine ; *k*, kidney ; *m*, mouth ; *n. c.*, nerve-cord ; *oe*, esophagus ; *o*, ovary ; *od*, oviduct ; *ph*, pharynx ; *r*, testes ; *s.r.*, seminal receptacles ; *v.v.*, ventral vessel.

each of which is modified for a particular part in the process. The mouth (*m*) leads into a muscular pharynx (*ph*) whose action enables the worm to retain its hold on various objects until swallowed, and this in turn is continuous with the esophagus. From here the food is passed into the thin-walled crop (*c*), and from this storehouse is gradually borne into the gizzard (*g*), whose muscular walls reduce it to a fine pulp now readily acted upon by the digestive fluids. These, resembling in their action the pancreatic juice of higher animals, are poured out from the walls of the intestine into which the food now makes its way; and as it courses down this relatively simple tube the nutritive substances are absorbed while the indigestible matters are cast away.

67. **Circulatory system.**—In all the groups of animals up to this point the digested food is carried through the body by a simple process of absorption, or in the threadworms by

means of the fluid in the body cavity; but in the earthworm the division of labor between different parts of the body is more perfect, and a definite blood system now acts as a distributing apparatus. This consists primarily of a dorsal vessel lying along the dorsal surface of the alimentary canal (Fig. 40), from which numerous branches are given off to the body wall, and to the digestive system through which they ramify in every direction before again being collected into a ventral vessel lying below the digestive tract. In some of the anterior segments a few of the connecting vessels are muscular and unbranched, and during life pulsate like so many hearts to force the blood over the body, forward in the dorsal vessel, through the " hearts " into the ventral vessel, thence into the dorsal by means of the small connecting branches.

Some of the duties of this vascular system are also shared by the fluid of the body cavity, which is made to circulate through openings in the partitions by the contractions of the body wall of the animal in the act of crawling. In this rough fashion a considerable amount of nutritive material and oxygen are distributed to various organs, and wastes are carried to the kidneys to be removed.

FIG. 41.—Diagram of earthworm kidney. *b*, blood-vessel ; *f*, funnel opening into body cavity; *o*, outer opening; *s*, septum; *w*, body wall.

68. **Excretion.**—In nearly all of the segmented worms there is a pair of kidneys to every segment (Figs. 40, 41). Each consists of a coiled tube wrapped in a mass of small blood-vessels, and at its inner end communicating with the body cavity by means of a funnel-shaped opening. In some unknown way the walls of the kidney extract the waste materials from the blood-vessels coursing over it and pass them into its tubular cavity. At the same time the cilia about the mouth of the funnel-shaped extremity are

driving a current from the body-cavity fluids, which wash the wastes to the exterior.

69. Nervous system.—The nervous system of the earthworm consists first of a brain composed of two pear-shaped masses united together above the pharynx (one shown in Fig. 34), from which nerves pass out to the upper lip and the head, which are thus rendered highly sensitive. Two other nerves also pass out from the brain, and, coursing down on each side of the pharynx like a collar, unite below it and extend side by side along the under surface of the digestive system throughout its entire extent. In each segment the two halves of this ventral nerve-cord are united by a nerve, and others are distributed to various organs, which are thus made to act and in proper amount for the good of the body as a whole.

In its relation to the outside world the chief source of information comes to the earthworm through the sense of touch, for definite organs of sight, taste, and smell are but feebly developed, while ears appear to be entirely absent. Nevertheless these are sufficient to enable it to lead a successful life, as is evidenced by the great number of such worms found on every hand.

70. Egg-laying.—In digging up the soil where earthworms abound one frequently finds small yellowish or brownish bodies looking something like a grain of wheat. These are the cocoons in which the earthworms lay their eggs, and the method by which this is performed is unique. We have already noted the presence of a swollen girdle (the *clitellum*) about the body of the worm. At the breeding season this throws out a fluid which soon hardens into an encircling band. By vigorous contractions of the body this horn-like collar is now slipped forward, and as it passes the openings of the reproductive organs the eggs and sperms are pushed within it. They thus occupy the space between the worm and the collar, and when the latter is shoved off over the head its ends close as though drawn to.

gether by elastic bands. A sac, the cocoon, is thus produced, containing the eggs and a milky, nutritive substance.
In a few weeks the worm
develops and, bursting the
wall of its prison, makes its
escape.

71. **Distribution.** — The
earthworms and their allies
are found widely distributed
throughout the world, and
all exhibit many of the
characters just described.
The greatest differences
arise in their mode of life :
some are truly earthworms,
but others are fitted for a
purely aquatic existence in
fresh water or along the
seacoast; a few have taken
up abodes in various ani-
mals and plants, and in
some of these situations they
extend far up the sides of
the higher mountains. In
all, the head is relatively
indistinct, the number of
bristles on each segment
few, and for this and other
reasons all are included in
the subclass Oligochæta, or " few-bristle " worms.

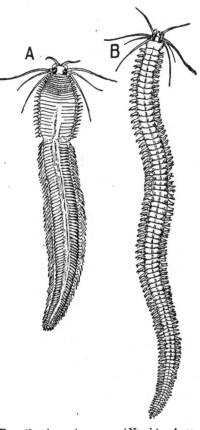

Fig. 42.—A marine worm (*Nereis*). A, appearance at breeding season, and B,
at other times.

72. **Nereis and its allies.**—In many of the above-mentioned situations members of a more extensive group of
worms are found, with highly developed heads and many
bristles arranged along-the sides of the body. These are
the Polychætes or " many-bristle " worms, and as a representative we may take *Nereis* (Fig. 42), a very common

form along almost any seashore. The body presents the
same segmented appearance as the earthworm, but the
head (Fig. 43, A) is provided with numerous sense organs,
chief among which are four eyes and
several tentacles or "feelers."

The segments behind the head

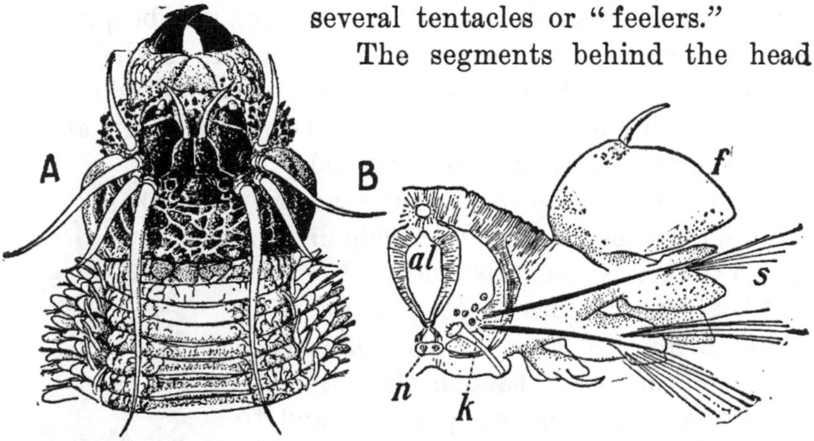

FIG. 43.—A, head and one of the lateral appendages (B) of a marine worm (*Nereis
brandtii*); *al*, intestine; *f*, "gill"; *k*, kidney; *n*, nerve cord; *s*, bristles for locomotion.

differ very little from one another, and, unlike those of
the earthworm, each bears a pair of lateral plates (Figs.
41, 42, B) or paddles with many lobes, some of which bear
numerous bristles. By a to-and-fro movement these organs
aid in pushing the animal about, or may enable certain species to swim with considerable rapidity.

As in all other worms, respiration takes place through
the surface of the body, the area of which is increased by
the development, on certain portions of the paddles (para-
podia), of plates penetrated with numerous blood-vessels,
which thus become special respiratory organs or gills
(Fig. 42, B).

In their internal organization the Polychætes are constructed practically on the same plan as the earthworms,
the principal difference being in the reproductive system.
In the earthworm this is restricted to some of the forward
segments, while in the present group the eggs and sperms

are developed in almost every segment, whence they are finally swept to the exterior through the tubes of the kidneys (Fig. 43, B).

The Nercis and its immediate relatives are all active forms, and by means of powerful jaws, which may be quickly extended from the lower part of the mouth cavity, they capture large numbers of small crustaceans, mollusks, and worms which happen in their path. Others more distantly related make their diet of seaweed, and many living on the sea bottom swallow great quantities of sand, from which they absorb the nutritious substances.

73. **Sedentary forms.**—Preyed upon by many enemies, a large number of species have been forced to abandon an active existence save in their early youth, and to construct many interesting devices for their protection. Numerous species, shortly after they commence to shift for themselves, build about their bodies tubes of lime (Fig. 45), from which they may emerge to gather food and into which they may dash in times of danger. As the worm grows the tube is correspondingly enlarged, and these tubes, in all stages of construction and variously coiled, may be found on almost every available spot at the seashore, and may often be seen on the shells of oysters in the markets.

In other species the tube is like thin horn, and may be further strengthened or concealed by numerous pebbles, bits of carefully selected seaweeds, or highly tinted shells, which give them a very attractive appearance. Such species usually develop out of immediate contact with other forms, but a few live so closely associated together that their twisted tubes

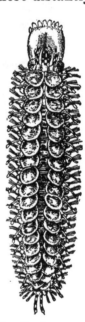

FIG 44.—A common marine worm (*Polynœ brevisetosa*), with extended proboscis and overlapping plates covering the back.

form great stony masses, sometimes several feet in diameter.

74. Effects of an inactive life.—In many species such a sedentary life has resulted in the almost complete disappearance of the lateral appendages, which therefore no longer serve as organs of respiration, and this function has been shifted accordingly on to other structures. These new organs are situated principally on the exposed head,

FIG. 45.—Sedentary tube-dwelling marine worms, upper left hand *Sabella* (one-half natural size), the remainder *Serpula* (enlarged twice). From life.

and Fig. 39 shows the general appearance of some common species. The corners of the mouth have expanded into great plumes, sometimes wondrously colored like a full-blown flower, and these, bounteously supplied with blood-vessels, act as gills. When disturbed, the plumes are hastily withdrawn into the tube, and some of the so-called serpulids (Fig. 45, bottom of figure) close the entrance with a funnel-shaped stopper. While the plumes are primarily respiratory organs, they also act as delicate feelers, and may even bear a score or more of eyes; and in addition, being covered with cilia, create the currents of water which bring minute organisms serving as food within reach of the mouth.

75. Development.—Unlike the earthworms, the Polychætes lay their eggs in the sea water, where they are left alone to develop as best they may. Both the male and female *Nereis*, as the egg-laying time approaches, undergo remarkable changes in their external appearance, resulting in the form shown in Fig. 42, A. They are now active swimmers, and thus are able to scatter the fertilized eggs over wide and more or less favorable areas. The young also for a time are free-swimming, but finally end their migrations by settling to the sea bottom, where they gradually attain the adult condition.

As in some of the flatworms, reproduction may also occur asexually by the division of the animal into two or more parts, each of which subsequently becomes a complete individual. In other species growth of various parts may result in two complete worms at the time of separation; and from such forms we may trace a fairly complete series up to those in which the original parent breaks up into twenty to thirty young.

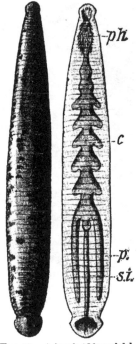

Fig. 46.—A leech (*Macrobdella*). Right-hand figure illustrates alimentary canal. *ph*, pharynx; *c*, crop; *p*, lateral pouches; *s.i.*, intestine.

76. The leeches.—At first sight the leeches (Fig. 46), or at least the smaller, more leaf-like forms, might be mistaken for flatworms, especially for some of the parasitic species. As in the latter, the mouth is surrounded by a sucker, and another is located at the hinder end of the body, but beyond this point the resemblance ceases. The outer surface is delicately marked off into eighty or a hundred rings, of which from three to five are included in one of the deeper true segments corresponding to those of other annelids. From two to ten pairs of simple eyes are borne on the head, and owing to the fact that they are active swimmers, or move by caterpillar-like looping, locomotor spines are unnecessary and absent. In their internal organization, however, there are many features which indicate a close relationship with the Oligochætes or few-bristle worms. The nervous, circulatory, and certain characteristics of the excretory systems are decidedly similar, but, on the other hand, there are some facts difficult to explain, which have led some zoologists to believe that the relationship of these animals can not at present be determined.

77. Haunts and habits.—The leeches usually dwell in among the plants in slowly running streams, but some occur in moist haunts on land, and a considerable number live in the sea. All are "bloodsuckers"—fierce carnivorous worms, whose bite is so insidiously made that the victim frequently is ignorant of their presence. Fishes, frogs, and turtles are the most frequently attacked, but cattle and other animals which come down to drink also become their prey. In some of the tropical countries the land-leeches are present in large numbers secreted among the leaves, and so severe are their attacks that various animals, even man, succumb to their united efforts. Adhering by their suckers, they puncture the skin, some using triple jaws, and fill themselves until they become greatly distended, when they usually drop off and digest the meal at leisure. In certain species the intestine is provided with lateral pouches (Fig. 45), which serve to store up the food until the time for digestion arrives. A full meal is sufficient with some species to last for two or three months, and the medicinal or horse-leech when gorged with food may consume a year in digesting it.

78. Egg-laying.—The eggs of some leeches are stored up in a cocoon like that of the earthworm, which is attached to submerged plants or placed under stones. When the young are able to lead independent lives they emerge with the form of the parent. A leaf-like form, *Clepsine*, sometimes found adhering to turtles, fastens the eggs to the under side of its body, and the young when hatched remain there for several days, adhering by their posterior suckers.

CHAPTER VII

ANIMALS OF UNCERTAIN RELATIONSHIPS

In this chapter we shall consider in a brief way a number of different groups of animals whose relationships are uncertain. Up to the present time the study of their habits, structure, and development has been of too fragmentary or unrelated a character to enable the majority of zoologists to agree upon their classification. Nevertheless, many of them are highly interesting and attractive, often very common, and in some respects they hold important positions in the animal kingdom.

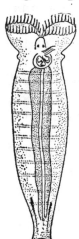

Fig. 47.—A wheel-animalcule (*Rotifer*).

79. The rotifers or wheel-animalcules.— The rotifers or wheel-animalcules are relatively small and beautiful organisms, rarely ever longer than a third of an inch, but at times so abundant that they may impart a reddish tinge to the water of the streams and ponds in which they live. At first sight they might be mistaken for one-celled animals, but the presence of a digestive tract and of reproductive elements soon dispels such a belief. Examined under the microscope, the more common forms are seen to possess an elongated body terminating at the forward end in two disk-like expansions beset along the edges with powerful cilia. These serve to drive the animal about, or, when it remains temporarily attached

83

by the sticky secretion of the foot, to sweep the food-particles down into the mouth. Through the walls of the transparent body such substances are seen to pass into the stomach, where they are rapidly hammered or rasped into a pulp by the action of several teeth located there. In the absence of a circulatory system the absorbed food is conveyed by the fluid of the body-cavity, which also conveys the wastes to the delicate kidneys. Several other features of their organization are of much interest, especially to the zoologist, who believes that he gains from their simple structure some ideas of the ancestors of the modern worms, mollusks, and their allies. During the summer the rotifers lay two sizes of "summer eggs," which are remarkable for developing without fertilization. The large size give rise to females, the smaller to males, the latter appearing when the conditions commence to be unfavorable. The "winter eggs," fertilized by the males and covered with a firm shell, are able for prolonged periods to withstand freezing, drought, or transportation by the wind. The adults also are able under the same adverse conditions to surround themselves with a firm protective membrane and to exist for at least a year. Once again in the presence of moisture the shell dissolves, and in a surprisingly short space of time they emerge, apparently none the worse for the prolonged period of quiescence.

80. Gephyrea.—There is a comparatively large group of worm-like organisms, over one hundred species in all, which at present hold a rather unsettled position in the animal kingdom. Some of the more common forms (Fig. 48) living in the cracks of rocks or buried in the sand, usually in shallow tide pools along the seashore, have a spindle-shaped body terminated at one end by a circlet of tentacles which surround the mouth. On account of their external resemblance to many of the sea-cucumbers (Fig. 95), they were earlier associated in the same group; but an examination of their internal organization inclines many zoologists

to the belief that the ancestors of some of these animals were segmented worms whose present condition has arisen possibly in accordance with their sluggish habits. This view is strengthened by the fact that in a very few species the larvæ are distinctly segmented, but lose this character in becoming adult. As before mentioned, the greater number of species live in burrows in the sand or crevices in the rocks, from which they reach out and gather in large quantities of sand. As these substances pass down the intestine the nutritive matters are digested and absorbed, while the indigestible matters are voided to the exterior. When large numbers are associated together they are doubtless important agents in modifying the character of the sea bottom, thus acting like the earthworms and their relatives.

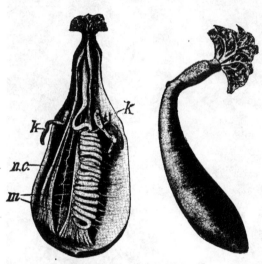

FIG. 48.—A gephyrean worm (*Dendrostoma*). Specimen on left opened to show *k*, kidney, *m*, muscle bands, and *n.c.*, nerve-cord.

81. **The sea-mats (Polyzoa).**—The sea-mats or Polyzoa constitute a very extensive group of animals common on the rocks and plants along the seashore, and frequently seen in similar situations in fresh-water streams. A few lead lives as solitary individuals, but in the greater number of species the original single animal branches many times, giving rise to extensive colonies. In some species these extend as low encrusting sheets over the objects on which they rest; while in others the branches extend into the

surrounding medium and assume feathery shapes (Fig. 49), which often bear so close a resemblance to certain plants

Fig. 49.—Lamp-shells or Brachiopods (on left of figure), fossil and living, and (on right) plant-like colonies of sea-mats.

that they are frequently preserved as such. What their exact position is in the animal scale it is somewhat difficult to say; but judging especially from their development, it appears probable that they are distant relatives of the segmented worms.

82. Lamp-shells or Brachiopods.—Occasionally one may find cast on the beach or entangled in the fishermen's lines or nets a curious bivalve animal similar to the form shown in Fig. 49. These are the Brachiopods, or lamp-shells. The remains of closely related forms are often abundant as fossils in the rocks (Fig. 49). Over a thousand species have been preserved in this way, and we know that in ages past they flourished in almost incredible numbers and were scattered widely over the earth. Unable to adapt themselves to changing conditions or unable to cope with their enemies, they have gradually become extinct, until to-day scarcely more than one hundred species are known. These are often of local distribution, and many are comparatively rare.

For a long period the Brachiopods, owing to their peculiar shells, were classed together with the clams and other bivalve mollusks. The presence of a mantle also strengthened the belief; but closer examination during more recent years has shown that the shells are dorsal and ventral, and not arranged against the sides of the animal as in the clams. Another peculiar structure consists of two great spirally coiled "arms," which are comparable in a general way to greatly expanded lips. The cilia on these create, in the water currents which sweep into the mouth, the small animals and plants that serve as food. The internal organization resembles in a broad way that of the animals considered in the previous section, and it now appears that both trace their ancestry back to the early segmented worms.

83. Band or nemertean worms.—In a few cases band or nemertean worms have been discovered in damp soil or in fresh-water streams. These are commonly small and inconspicuous, and are pigmies when compared with their marine relatives, which sometimes reach a length of from fifty to eighty feet. Many of the marine species (Fig. 50) are often found on the seashore under rocks that have been exposed by the retreating tide. They are usually highly colored with yellow, green, violet, or various shades of red, and are so twisted into tangled masses that the different parts of the body are indistinguishable. As the animal crawls about, a long thread-like appendage, the proboscis, is frequently shot out from its sheath at the forward end of the body and appears to be used as a blind man uses his stick. At other times, when small worms and other animals are

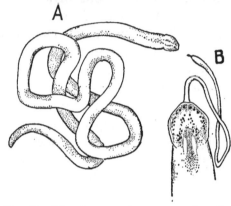

FIG. 50.—A band or nemertean worm. A, entire worm; B, head, bearing numerous eyes and spine-tipped proboscis.

encountered, the proboscis is shot out farther and with greater force, impaling the victim on a sharp terminal spine (Fig. 50). The food is now borne to the mouth, located near the base of the proboscis, is passed into the digestive tract, traversing the entire length of the body, and is farther operated on by systems of organs too complex to be considered here.

CHAPTER VIII

MOLLUSKS

84. General characters.—For very many years the mollusks—that is, the clams, snails, cuttlefishes, and their allies—have been favorite objects of study largely because of the durability, grace, and coloration of the shell. The latter may be univalve, consisting of one piece, as in the snails, or bivalve, as in the clams and mussels, and may possess almost every conceivable shape, and vary in size from a grain of rice to those of the giant clam (Tridacna) of the East Indian seas, which sometimes weighs five hundred pounds. These external differences are but the expression of many internal modifications, which, while adapting these animals for different modes of life, are yet not sufficient to disguise a more fundamental resemblance which exists throughout the group. In some respects the mollusks show a close resemblance to the annelid worms, but, on the other hand, the body is usually more thick-set and totally devoid of any signs of segmentation. In every case the skin is soft and slimy, demanding moist haunts and usually the protection of a shell, and the body is modified along one surface to form a foot or creeping disk which serves in locomotion. The internal organization is somewhat uniform, and will admit of a general description later on. Mollusks are divided into three classes, viz.: The Lamellibranchs, embracing the clams; the Gasteropods, or snails; and the Cephalopods, or cuttlefishes, squids, and related forms.

85. Lamellibranchs (clams and mussels).—Numerous representatives of this class, such as the clams and mussels, occur along our seacoasts or are plentifully distributed in the fresh-water streams and lakes. They are distinguished from other mollusks by a greatly compressed body, which is enclosed in a shell consisting of two pieces or *valves* locked together by a hinge along the dorsal surface. Raising one of these valves, the main part of the body may be seen to occupy almost completely the upper (dorsal) part of the shell (Fig. 51), and to be continued below into the muscular hatchet-shaped foot (*ft.*), which aids the clam in plowing its way through the sand or mud in which it lives. Arising on each side of the back of the animal and extending its entire length is a great fold of skin, which completely lines the inner surface of the corresponding valve of the shell. These are the two mantle lobes (*m*) instrumental in the formation of the shell, and enclosing between them a space containing the foot and a number of other important structures, the most conspicuous of which are the gills (*g*), consisting of two broad, thin plates attached along the sides of the animal and hanging freely into the space (mantle cavity) between the mantle and the foot. Owing to this lamella-like character of the branchia or gills the class derives its name, lamellibranch. To illustrate the relations of these various organs to one another the clam has been compared to a book, in which the shells are represented by the cover, the fly-leaves by the mantle lobes, the first two and last two pages by the gills, and the remaining leaves by the foot. In the clams, however, the halves of the mantle, like the halves of the shell, are curved, and thus enclose a space, the mantle cavity, which is partly filled by the gills and foot.

Unlike the other mollusks which usually lead active and more aggressive lives, the clams show scarcely a sign of a head and tentacles, and other sense organs are likewise absent from this region. The mouth also lacks definite organs of mastication, and as devices for catching and holding food are not developed, the food is brought to the mouth by means of the cilia on the great triangular lips or *palps* which bound it on each side (Fig. 51, A, *p*).

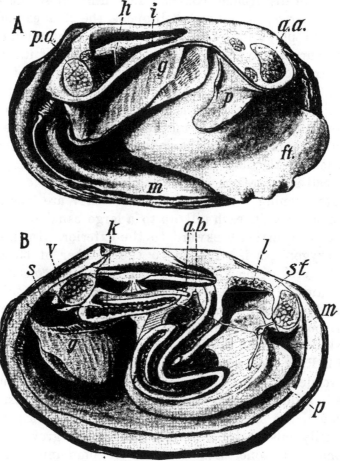

FIG. 51.—Anatomy of fresh-water clam. A, right valve of shell removed ; B, dissection to show internal organs. *a*, external opening of kidney ; *a.a.*, the anterior muscle for closing the shell ; *b*, opening of reproductive kidney ; *c*, brain ; *ft.*, foot ; *g*, gill ; *h*, heart ; *i*, intestine ; *k*, kidney ; *l*, liver ; *m*, mantle (upper fig.), mouth (lower fig.) ; *p*, palp (upper fig.), foot nerves (lower fig.) ; *p.a.*, hinder muscle for closing the shell ; *s*, space through which the water passes on leaving the body ; *st*, stomach ; *v*, nerves supplying viscera.

Between the halves of the shell in the hinge region is a horny pad that acts like a spring, and without any muscular effort on the part of the clam keeps the shells open.

These are also united by two great adductor muscles, located at opposite ends of the animal (Fig. 51, A, *a.a.*, *p.a.*), which in times of disturbance contract and firmly close the shell. Upon their relaxation the shell opens, the clam extends its foot, and plows its way leisurely through the mud, or remains buried, leaving only the hinder portion of its gaping shell exposed. Through this opening a current of water is continually passing in and out, owing to the action of the cilia covering the gills, and by placing a little carmine or coloring matter in the ingoing stream we may trace its course through the body. Passing in between the mantle and the foot it travels on toward the head, giving off small side streams which are continually made to enter minute openings in the gills, whence they are conducted through tubes in each gill up to a large canal at its base, where it is carried backward to the exterior. In this process oxygen gas is supplied to the number of blood-vessels traversing the gills, and at the same time considerable quantities of minute organisms and organic *débris* are hurried forward toward the head, where they encounter the whirlpools made by the cilia on the lips and are rapidly whisked down into the mouth and swallowed.

86. **Rock- and wood-boring clams.**—Other similar forms are rendered even more secure through their ability to bore in solid rock. In the common Piddock, for example (Fig. 52), the shell is beset with teeth like a rasp, which gradually enlarge the cavity as the animal grows, until it becomes a prisoner with no means of communication with the exterior save the small opening through which the siphons project. This is also the ease with the *Teredo*, frequently called the shipworm, which swims about for some time during early life and then, about the size of a small pinhead, settles down upon the timbers of wharves or unsheathed ships, into which it rapidly tunnels. Throughout life its excavation is extended sometimes to a distance of two to three feet, and imprisoned yet safe **at**

the bottom of its burrow, it extends its slender siphons up the tube and out of the entrance for its food supply. Often hundreds of individuals enter the same piece of wood, which becomes thoroughly riddled within a short

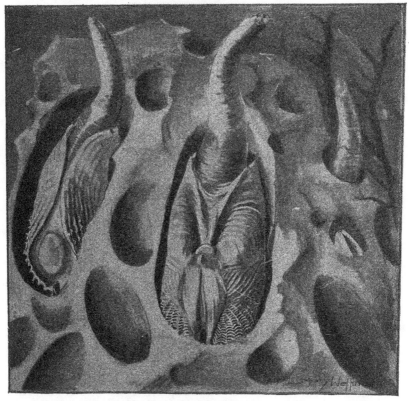

Fig. 52.—The piddock (*Zirphæa crispata*), a rock-boring mollusk. Natural size, from life.

time, and though giving no outward sign of weakness may collapse with its own weight. Incalculable damage is thus rendered to the shipping interests, and in consequence much has been done to check their ravages, but they are far from being completely overcome.

87. Other stationary species.—A large number of other species, while small and inconspicuous, are also free to

move about, but as they become larger they lose this ability either wholly or periodically. In the edible mussels (*Myti-lus*, Fig. 53), for example, which are associated in great numbers on the rocks along our coasts, the foot early becomes long and slender and capable of reaching out a considerable distance from the shell to attach threads (byssus), which it spins, to foreign objects. These are remarkably strong, and when several have been spun it becomes a matter of much difficulty to dislodge them. After remaining anchored in one situation for a while the mussel may vol-

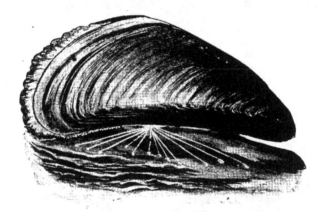

Fig. 53.—The edible mussel (*Mytilus edulis*), showing the threads by which it is attached. Natural size, from life.

untarily free itself, and in a labored fashion move to some other more favorable spot where it again becomes attached, but there are numerous species, such as "fan shells" (*Pinna*), scallops, *Anomia*, and a few fresh-water forms, where the union is permanent.

Finally, in the oysters, some of the scallops, and a number of less familiar forms, the young in very early life drop down upon some foreign object to which the shell soon becomes firmly attached, and in this same spot they pass the remainder of their lives. The oyster usually falls upon the left half of its shell, which becomes deep and capacious enough to contain the body, while the smaller right valve acts as a lid. As locomotion is out of the question, the foot never develops, and the shell is held by only one adductor muscle, whose point of attachment in the oyster is indicated by a brown scar in the interior of the shell.

88. **Internal organization.**—It is thus seen that the external features of the clam are variously modified, according to the life of the animal, but the internal organization is much more uniform. In nearly every species the food consists of floating organisms, which are driven by the palps into the mouth and on to the simple stomach, where it is subjected to the solvent action of the fluids from the liver (Fig. 51, B, *l*) before entering the intestine. This latter structure is usually of considerable length, and in the active species extends down into the foot, and it is also peculiar in passing through the ventricle of the heart. Traversing the intestine the nutritive portion of the food is absorbed, and is conveyed over the body by a circulatory system more highly developed than in the higher worms. On the dorsal side of the clam, in a spacious pericardial chamber, the large heart is situated (Fig. 51, *h*), consisting of a median highly muscular ventricle surrounding the intestine and of two thin auricles, one on either side. From the former, two arteries with their numerous branches convey the blood to all parts of the body, where it accumulates, not in capillaries and veins, but in spaces or *sinuses* among the muscles and various organs, constituting a somewhat indefinite system of channels which lead to the gills and kidneys. In these organs the blood delivers up the waste which it has accumulated on its journey, and absorbing a supply of oxygen, it flows into the great auricles, which in turn pass it into the ventricle to circulate once more throughout the body.

The excretory apparatus, consisting usually of two kidneys, of which one may degenerate in many snails, bears a close resemblance to that of the annelids. In the clam, for instance, each consists of a bent tube symmetrically arranged on each side of the body (Fig. 51, B, k), and the inner ends (a), corresponding to the ciliated funnel of the anne- iid kidney, open into the pericardial cavity. The walls are continually active in extracting wastes from the blood supplied to them, and these, together with the substances swept out from the pericardial cavity, traverse the tube and are carried to the exterior. In other mollusks the kidney may be more compact, or greatly elongated, or otherwise peculiar, but in reality they bear a close resemblance to those of the clam.

89. **Nervous system.**—The nervous system, like the ex- cretory, differs considerably in different mollusks, yet the resemblances are fairly close throughout. In the clam the cerebral ganglia corresponding to the "brain" in annelids is located at either side, or above the mouth, and from it several nerves arise, the larger passing downward to two pedal ganglia (p) embedded in the foot and to the visceral ganglia (v) far back in the body (Fig. 51, B). These nerve centers continually send out impulses which regulate the various activities of the body and also receive impressions from without. These come chiefly through the sense of touch, for in the clams the other senses are usually either feebly developed or altogether absent.

90. **Development.**—In the mollusca new individuals al- ways arise from eggs, which are commonly deposited in the water and there undergo development. In the fresh-water clams the reproductive organ is usually situated in the foot (Fig. 51), while in the oyster and similar inactive species it is attached to the large adductor muscle. In these latter, and in many other marine forms, the eggs are shed directly into the sea, where they are left to undergo their development buffeted by winds and waves and subject to the attack of numerous enemies. Under such circumstances the chances of survival are slight, and for this reason eggs are laid in vast numbers, which have been variously estimated for the oyster, for example, from two to forty million. Develop-

ment proceeds at first much as in the sponge, but soon the shell, foot, gills, and various other molluscan structures put in an appearance, and the few surviving young which have been free-swimming now settle down in some favorable spot, and attach themselves or burrow according to their habit.

91. **Life history of fresh-water clams.**—The life history of our common fresh-water clams is perhaps one of the most remarkable known among mollusks. The parent stores the eggs, as soon as they are laid, in the outer gill plate, and there, well protected, they undergo the first stages of their development, which results in the formation of minute young enclosed in a bivalve shell beset with teeth. These are often readily obtained, sometimes as they are escaping from the parent, and when examined under the microscope are seen to rapidly open and close their shells in a snapping fashion when in the least disturbed. In a state of nature this latter movement may result in attaching the young to the fins or gills of some passing fish, which is necessary to its further development. Within a short time it becomes completely embedded in the flesh of its host, from which, as a parasite, it draws its nourishment, and during the next few weeks undergoes a wonderful series of transformations resulting in a small mussel, which breaks its way through the thin skin of the fish and drops to the bottom.

92. **The gasteropods.**—The gasteropods, including snails, slugs, limpets, and a host of related forms, fully twenty thousand different species in all, are found in most of our fresh-water streams and lakes and in moist situations on land, while great numbers live along the seashore and at various depths in the ocean, even down as far as three miles. Examining any of them carefully we find many of the same organs as in the clams, but curiously changed and adapted for a very different and usually active life. In our common land snails (Fig. 54), which we may well examine before passing on to a general survey of the group, the first

striking peculiarity is in the univalve shell, with numerous whorls, into which the animal may at any time withdraw completely. Ordinarily this is carried on the back of the spindle-shaped body, which is fashioned beneath into a great

Fig. 54.—The slug (*Ariolimax*) and common snail (*Helix*). From life.

flat sole or creeping surface that bears on its forward border a wide opening through which mucus is continually issuing to enable the snail to slip along more readily. Slime also exudes on other points on the surface of the body and affords a valuable protection against excessive heat and drought.

Unlike the clams, the forward end of the body is developed into a well-marked head bearing the mouth and a complicated mechanism for gathering and masticating food, together with two pairs of tentacles, one of which carries the eyes. On the right side of the animal, some distance behind the head, is the opening of the little sac-like mantle cavity (Fig. 54) which contains the respiratory organs, and into which the alimentary canal and the kidneys pour their wastes. The relation of these organs to the mantle cavity is the same as in the clams, though the cavities differ much in size and position.

93. **Other snails. The shell.**—Extending our acquaintance to other species of snails, we find the same general plan of body, although somewhat obscured at times by

many modifications. A foot is generally present, also a more or less well-developed head, and the body is usually surrounded by a shell which varies widely in shape and size in different species. In the common limpets the early coiled shell is transformed into an uncoiled cap-like one, and in the keyhole limpets is perforated at its summit. The chitons or armadillo-snails (Fig. 55), often found associated with the limpets, carry a most peculiar shell consisting of eight plates, which enables the animal to roll up like an armadillo when disturbed. A shell is by no means a necessity, however, for in many species, such as the beautiful naked snails or Nudibranchs (Fig. 56) common along our coasts, it may be entirely absent, or, as in the ordinary slugs, reduced to a small scale embedded in the skin.

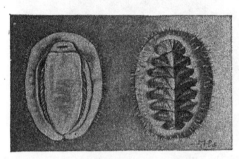

Fig. 55.—The chiton, armadillo-snail or sea-cradle. The left-hand figure shows mouth in center of proboscis, the broad foot on each side of which are numerous small gills. The right-hand figure shows the mantle and shell, composed of eight plates. From life, one-half natural size.

94. Respiration.—A considerable quantity of oxygen is absorbed through the skin, as in all mollusks, but the chief part of the process is usually taken by the plume-like gills, one or two in number, which are located in the mantle cavity. In the chitons (Fig. 55) the number of gills is greater, amounting in some species to over a hundred, while in the Nudibranchs (Fig. 56) gills are absent, their places being taken by more or less feathery expansions of the skin on the dorsal surface.

Many of the gasteropods left exposed on the rocks by a retreating tide retain water in the mantle cavity, from which they extract the oxygen until submerged again.

Others breathe by means of gills while under water, and by
the surface of the body and the moist walls of the mantle

Fig. 56.—Three different species of naked marine snails or Nudibranchs. Natural
size, from life.

cavity when exposed.   In some of the small Littorinas
attached so far from the sea as to be only occasionally
washed by the surf this latter method may prevail for days
together—in fact they live better out of water than in it.
It is not difficult to imagine that such forms, keeping in
moist places, might wander far from the sea, and, losing
their gills, become adapted to a terrestrial life.   It is
believed that in past times this has actually occurred, and
that our land forms trace their descent from aquatic ancestors.   To-day they breathe by a lung—that is, they take
oxygen through the walls of the mantle cavity, as the slug
may be seen to do, though in some species traces of the old
gill yet remain.

95. **Food and digestive system.**—Many mollusks live upon
seaweeds, and the greater number of terrestrial forms are
fond of garden vegetables or certain kinds of lichens, but,
on the other hand, the latter, together with a large number
of marine snails, are carnivorous.   In all cases the food
requires to be masticated, and, unlike the clams, the mouth
is usually provided with horny jaws, and an additional

masticatory apparatus which consists of a kind of tongue
with eight to forty thousand minute teeth in our land
forms (Fig. 57), while in certain marine snails they are
beyond computation. With the licking motion of the
tongue this rasp tears the food into shreds before it is
swallowed, and in the whelks or borers it serves to wear a
circular hole through the shells of other mollusks, which
are thus killed and devoured.
This latter process is facilitated by the secretion of the
salivary glands, which has a
softening effect upon the
shell. Ordinarily the saliva
of snails exercises some digestive action.

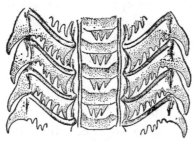

FIG. 57.—A small portion of the radula or
tongue-rasp of a snail (*Sycotypus*).

In the stomach of some
snails are teeth or horny
ridges which also are instrumental in crushing the food,
and in numerous minor respects peculiarities exist in different species according to the nature of the food; but in its
general features the digestive tract is similar to that of
the clams.

The processes of circulation and excretion are also carried on by means of systems which show a certain resemblance to those of the clams. As might be expected, certain
differences exist, sometimes very great, but they are of too
technical a nature to concern us further.

96. **Sense-organs of lamellibranchs and gasteropods.**—
The eyes of mollusks differ widely in their structure and
the position they occupy in the body. In our common
land snails two pairs of tentacles are borne on the head,
the lower acting as feelers, while each of the upper ones
bears on its extremity the eye, appearing as a minute black
dot (Fig. 54). In this same position the eyes of many
marine snails occur, but there are numerous species in
which there are other accessory eyes. In many of the

limpets, for instance, there are numbers of additional eyes
carried on the mantle edge just under the eaves of the
shell, and forming a row completely encircling the body.
(In the scallops there are two rows of brilliantly colored
eyes, set like jewels on the edges of the mantle just within
the halves of the shell.) In the chitons the eyes of the
head disappear by the time the animal attains maturity,
and in some species at least their place appears to be taken
by great numbers of eyes, sometimes thousands, which are
embedded in the shells. On the other hand, eyes are completely absent in certain species of burrowing snails and in
several living in the gloomy depths of the sea far from the
surface; they appear to be absent also from fresh-water
clams; but the fact that certain species close their shell
when a shadow falls upon them, leads to the belief that
while actual eyes are not present the skin is extremely
sensitive to light. This is also the case with many snails.

97. **Smell.**—Since the sense of sight is generally undeveloped in the mollusks, they rely chiefly upon touch and
smell for recognizing the presence of enemies and food.
Tentacles upon the head and other parts of the body, and
a skin abundantly supplied with nerves, show them to possess a high degree of sensibility; but in the greater number of species the sense of smell is of chief importance.
Many experiments show that tainted meat and strongly
scented vegetables concealed from sight and several feet
distant from many of our land and sea mollusks will attract
them at once. In these forms the sense of smell appears to
be located on the tentacles, but additional organs, possibly
of smell, are located on various portions of the body, usually in the neighborhood of the gills.

98. **Taste and** hearing.—Several mollusks appear to be
almost omnivorous, but others are decidedly particular in
their choice of food, which leads us to suspect that they
possess to some extent the sense of taste. Nerves supplying the base of the mouth have also been detected, which

may be those of taste; but experiments along the line are difficult to perform, and our knowledge of this subject is far from complete. The same is true of hearing. Certain organs, interpreted as ears and located in the foot, have the form of two hollow sacs, containing one or more solid particles of sand or lime, whose jarrings, when effected by sonorous bodies, may result in hearing. On the other hand, it is held by some that they, like the semicircular canals of higher animals, may regulate the muscular movements which enable the animal to keep its balance.

99. **Egg-laying habits and** development.—The egg-laying habits of the gasteropods differ almost as widely as their haunts. The terrestrial forms lay comparatively few eggs, ranging in size from small shot to a pigeon's egg in some of the tropical species. These are buried in hollows in the ground or under sticks and stones, and after a few weeks hatch out young snails having the form of the adult. The same is also true of most of the fresh-water snails, which lay relatively smaller eggs embedded in a gelatinous mass frequently found attached to sticks and leaves, or on the walls of aquaria in which they are confined. Many marine species construct capsules of the most varied patterns which they attach to different objects, and in these the young are protected until they hatch. In the limpets and many of the chitons the eggs are laid by thousands directly in the water, and after a short time develop into free-swimming young, differing considerably from the parent in appearance. Those escaping the ravages of numerous enemies finally settle down in a favorable situation and gradually assume the form of the adult.

100. **Age, enemies, and means of defense of lamellibranchs and gasteropods.**—How much time is consumed by the young in growing up, and the length of time they live, are questions generally unsettled. It is said that the oyster requires five years to attain maturity, and lives ten years; the fresh-water clam develops in five years, and some species live from

twelve to thirty years; and the average length of life of the snail appears to be from two to five years. Certain it is that mollusks have numerous enemies besides man which prevent multitudes from living lives of normal length. Birds, fishes, frogs, and starfishes beset them continually, and many fall a prey to the ravages of internal parasites or to other mollusks. Under ordinary circumstances the shell is sufficient protection, and the spines disposed on the surface in many species render the occupant still less liable to attack. Many snails carry on the foot a horny or calcareous plate known as the operculum, which closes the entrance of the shell like a door against intruders. Certain noxious secretions poured out from the skin also serve as a means of defense, and many Nudibranchs (Fig. 56) bear nettle-cells on the processes of the body, which probably render them distasteful to many animals. Finally, there are numerous clams, mussels, snails, and slugs whose colors harmonize so closely with their surroundings that they almost completely baffle detection, and enable them to lead as successful a life as those provided with special organs of defense.

101. **Cephalopods.**—The animals belonging to this class, such as the squids and cuttlefishes (Fig. 58), are by far the most highly developed mollusks. They are of great strength, capable of very rapid movements, and several species are many times the largest invertebrates. In almost every case there is a well-defined head bearing remarkably perfect eyes, and also a circle of powerful arms provided with numerous suckers which aid in the capture of food (Fig. 58). Posteriorly the body is developed into a pointed or rounded visceral mass which to a certain extent is free from the head, giving rise to a well-marked neck. Some forms, such as the squids (Fig. 58, upper figure), are provided with fins which drive the animal forward, but in common with other cephalopods they are capable of a very rapid backward motion. By muscular movements water is taken

into the large mantle cavity within the body, a set of valves prevents its exit through the same channels, and upon a vigorous contraction of the body walls the water is forced out rapidly through the small opening of the funnel, which

FIG. 58.--Cephalopods. Lower figure, the devil-fish or octopus (*Octopus punctatus*). The upper figure represents the squid (*Loligo pealii*) swimming backward by driving a stream of water through the small tube slightly beneath the eyes. From life, one-third natural size.

drives the animal backward after the fashion of an exploding sky-rocket. In this way they usually escape the fishes and whales that prey upon them, but an additional device has been provided in the form of a sac within the body, whose inky contents may be liberated in such quantity as to cloud the water for a considerable distance, and thus enable them to slip away unseen into some place of safety.

Most of the cephalopods are further protected by their ability to assume, like the chameleon, the color of the object upon which they rest. In the skin are embedded multitudes of small spherical sacs filled with pigments of various colors, chiefly shades of red, brown, and blue, each sac being connected with a nerve and a series of delicate muscles. If the animal settles upon a red surface, for example, a nerve impulse is sent to each of the hundreds of color sacs of corresponding shade, causing the muscles to contract and flatten the bag like a coin, and thus exposing a far greater surface than before, they give the animal a reddish hue. In the twinkling of an eye they may completely change to another tint, or present a mottled look, and some may even throw the surface of the skin into numerous small projections that make the animal appear part of the rock upon which it rests. These devices not only serve for protection, but they also aid in enabling these mollusks to steal upon their prey, chiefly fishes, which they destroy in great numbers with lionlike ferocity.

The devil-fishes and a number of other species are usually found creeping along the sea bottom, generally near shore, and are solitary in their habits, while the squids remain near the surface and frequently travel in great companies, sometimes numbering hundreds of thousands. In size they usually range from a few inches to a foot or two in length, but a few devil-fishes and squids attain a greater size, some of the latter reaching the enormous length of from forty to sixty feet. There are many stories of their great strength and of their voluntarily attacking people and even overturning boats, but the latter are in almost every case sailors' yarns.

In their external organization the cephalopods have little to remind one of any of the preceding mollusks, and their internal structure shows only a distant resemblance. In the Octopi (Fig. 58) the shell is lacking; in the squid it is called the pen, and consists of a horn-like substance without any lime deposit; in the cuttlefishes it is spongy and plate-like, and is a familiar object in the shops; and, finally,

in the nautilus it is coiled and of considerable size, and, unlike that of any other cephalopod, it is carried on the outside of the animal. Interiorly it is divided by a number of
partitions into chambers, the last one of which is occupied
by the animal.

The alimentary canal shows some resemblance to that
of other mollusks, but, as in the case of the other systems
of the body, it possesses a far higher state of development.
The mouth is situated in the center of a circle of arms,
which in reality are modified portions of the foot, and is
furnished with two parrot-like jaws. From this point the
esophagus leads back into the body mass to the stomach,
which with the liver and intestine are sufficiently like
those of the clam and snail to require no further comment.

Respiration is effected by the skin to a certain extent,
but chiefly by two gills (four in the nautilus), and the circulatory system, which conveys the blood to and from these
organs and over the body with its complex heart, arteries,
capillaries, and veins, is more highly developed than in
any other invertebrate.

As might be expected in animals with so great sagacity
and cunning, the nervous system and the sense-organs reach
a degree of development but little short of what we find in
some of the vertebrates. The chief part of the nervous
system is located in the head, protected by a cartilaginous
skull, a very rare structure among invertebrates; and while
the different ganglia may be recognized in a general way
and be found to correspond to a certain extent to those
of foregoing mollusks, they are so largely developed and
massed together that it is impossible at present to understand them fully. From this point nerves pass to all
regions of the body, to the powerful muscles, the viscera,
and the organs of special sense, controlling the complex
mechanism in all its workings.

There is no doubt that the cephalopods see distinctly
for considerable distances, and a careful examination of

the eye of the squids and cuttlefishes has shown them to be remarkably complex and in many respects to be constructed upon much the same plan as those of the vertebrates. As to the other senses not so much is known, but undoubtedly many species of cephalopods are possessed of a shrewdness and cunning not shared by any other invertebrates, save some of the insects and spiders, and are vastly more highly organized than their molluscan relatives.

CHAPTER IX

THE ARTHROPODS

102. General characters.—In the Arthropods, that is, **the** crabs, lobsters, shrimps, insects, spiders, and a vast host of related forms, the body is bilaterally symmetrical, and is composed of a number of segments arranged in a series, as in the earthworm and other annelids. A hornlike cuticle, sometimes called the shell, bounds the external surface—in early life thin and delicate, but later relatively thick, and often further strengthened by lime salts. Along the line between the segments this coat of mail remains thin and forms a flexible joint. Appendages also are borne on each segment, not comparatively short and fleshy outgrowths like the lateral appendages of many of the worms, but usually long and jointed (hence the name Arthropod, meaning jointed foot), and variously modified for many different uses.

103. Classification.—The species belonging to this group outnumber the remainder of the animal kingdom. Their haunts also are most diverse. Some are adapted for lives in the sea and fresh water, others for widely different situations on land, and a great number are constructed for a life on the wing. A certain resemblance exists among them all, but the modifications which fit them for their different habitats are also profound, and have resulted in the division of the Arthropods into five classes. The first class (*Crustacea*) contains the crayfish, crabs, etc.; the second (*Onychophora*) includes the curious worm-like peripatus (Fig.

109

72); the third (*Myriapoda*, meaning myriad-footed) embraces the centipeds and "thousand-legs"; the fourth (*Insecta*) contains the insects; and the fifth (*Arachnida*) includes the scorpions, spiders, and mites.

104. **The Crustacea.**—The number of species of crustaceans is estimated to be about ten thousand, and while the greater number of these are marine, many are found in fresh water and a few occur on land. in size they range from almost microscopic forms to the giant crabs and lobsters. They differ also in shape to a remarkable degree, but at the same time there is a decided resemblance throughout the group, except in those species which have become modified by a parasitic habit. The characteristic external skeleton is invariably present, and gives evidence of the deep internal segmentation of the body. In the simple Crustacea this is very apparent, but in the higher forms it is usually more or less obscured, owing to the fusion of some of the different segments, especially those of the head, as in the crayfish (Fig. 65).

The class of the Crustacea is subdivided into two subclasses (*Entomostraca* and *Malacostraca*), the first containing the fairy-shrimps (*Branchipus*, Fig. 59) and their allies, the copepods (such as Fig. 60), the barnacles (Fig. 61), and a number of other species. In their organization all are comparatively simple, usually small, and the appendages show relatively little specialization. The other subclass contains the more highly developed and usually large-sized Crustacea, among which are the shrimps, crayfishes, lobsters, crabs, and a number of other forms.

105. **Some simple Crustacea.**—While the members of the first subclass are minute and inconspicuous, several species are often remarkably abundant in our small fresh-water pools. Among these is the beautifully colored fairy-shrimp (*Branchipus*, Fig. 59), with greatly elongated body and leaf-like appendages, whose relatively simple character leads the zoologist to think that they are among the simplest

Crustacea, and in several points resemble the ancestral form from which all the modern species have descended. Some nearly related forms are provided with a great fold of the body-wall, which may almost completely conceal the animal from above, or it may be formed like a bivalve clam-shell, within which the entire body may be withdrawn. This

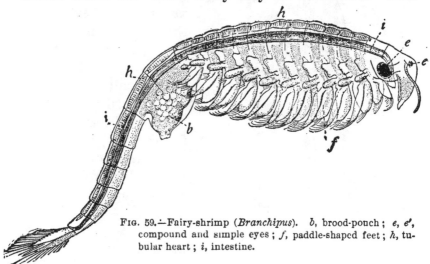

Fig. 59.—Fairy-shrimp (*Branchipus*). *b*, brood-pouch; *e, e'*, compound and simple eyes; *f*, paddle-shaped feet; *h*, tubular heart; *i*, intestine.

latter character is also found in the water-fleas (*Daphnia*), very much smaller forms, and sometimes occurring in millions on the bottoms of our ponds and marshes. They are readily distinguished from the fairy-shrimp by the shortness of the body, the small number of appendages, and by their habit of using their antennæ as swimming organs, which gives to their locomotion a jerky, awkward character.

106. **Cyclops and relatives.**—*Cyclops* (Fig. 60), the representative of a number of lowly forms belonging to the order of Copepods, is one of the commonest fresh-water Crustacea. The forward segments of the spindle-shaped body are covered by a large shield or carapace, the feet are few in number, and, like its fabled namesake, it bears an eye in the center of the forehead. Nearly related species are also remarkably abundant at the surface of the sea, at times occur-

ring in such vast numbers that they impart a reddish tinge to the water over wide areas, and at night are largely responsible for its phosphorescence. Many others are parasitic in their habits, and scarcely a salt-water fish exists but that at one time or another suffers from their attacks. On the other hand, many fresh- and salt-water fishes depend upon the free-swimming forms for food, and hence, from an economic point of view, they are highly important organisms.

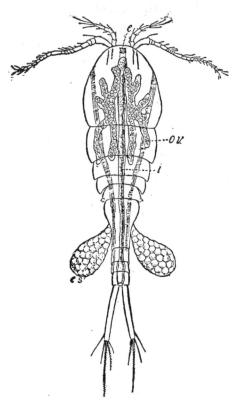

Fig. 60.—Cyclops.   *e. s.*, eggs; *i*, intestine; *ov*, reproductive organ.

107. Barnacles.—The parasitic habit and the lack of locomotion has also produced marvelous changes among the barnacles, so great that originally they were placed among the mollusks; and as with the parasitic copepods, their true position was only known after their life-history had been determined. In the goose-barnacles * the body, attached by a fleshy stalk to foreign objects, is enclosed by a tough membrane, corresponding to the carapace of other Crustacea, in which are embedded five calcareous plates. This

* So called because of the belief, which existed for three hundred years prior to the present century, that when mature these animals give birth to geese.

is open along one side, and allows the feather-like feet to
project and produce currents in the surrounding water
which brings food within reach. In the acorn-barnacles
(Fig. 61) the stalk is absent, and the body, though possess-

Fig. 61.—Barnacles. Acorn-barnacles chiefly in lower part of figure ; goose-barnacles
above. Natural size.

ing the same general character as the goose-barnacles, is
shorter, and enclosed in a strong palisade consisting of six
calcareous plates.

The larger number of barnacles attach themselves to
the supports of wharves, the hulls of ships, floating timbers, the rocks from the shore-line down to considerable
depth, and a few species occur on the skin of sharks and
whales. On the other hand, there are several species which
are parasitic, and in accordance with this mode of life exhibit various degrees of degeneration. In the most extreme

cases (*Sacculina*) the sac-like body, attached to the abdomen of crabs, is entirely devoid of appendages and any signs of segmentation. A root-like system of delicate filaments extends from the exposed part of the animal into the host and absorbs the necessary nutriment. The mouth and alimentary canal are accordingly absent—in fact, the body contains little but the reproductive organs and a very simple nervous system.

108. **Structure.**—In the internal organization of these smaller crustaceans many differences may be noted, though they are usually less profound than the external. Ordinarily the alimentary canal is a straight tube passing through the body, and is provided with a pouch-like stomach, and a more or less clearly defined liver. In all, except the parasitic species, the external mouth-appendages masticate the food, and in a very few of the above-described groups it may be further ground between the horny ridges on the stomach-walls. After this preliminary treatment it is subjected to the action of the digestive juices, and when liquefied is absorbed into the body. Here it is circulated by a blood-system of widely different character. In many cases definite arteries and veins are absent. The blood courses through the body in the spaces between the different organs propelled by the beating of the heart, which it is made to traverse. In Cyclops (Fig. 60) even the heart is absent, and the blood is made to circulate by contractions of the intestine. In most of these smaller Crustacea considerable oxygen is absorbed through the body-wall; but in several species, for example, the fairy-shrimp (Fig. 59), special gills are developed on the appendages of the body.

109. **Multiplication.**—Among the Crustacea thus far considered the males are usually readily recognized owing to their small size. The females also are usually provided with brood-pouches in which the developing eggs are protected. In almost every case the young are born in the

form of minute larvæ, provided with three pairs of appendages, a median eye (Fig. 62), and a firm external skeleton
or cuticle. This latter prevents the continuous growth of
the larvæ or *nauplius*, and every few days it is thrown off,
and while the new one is forming the body enlarges. During this time new appendages are developed, so that after

each moult the young crustacean emerges less like its
former self and more and more
like its parents. In the barnacles, after several moults
have taken place, the young
become permanently attached
by means of their first antennæ, their thoracic feet change
into feathery appendages, and
several other changes occur.
In some of the parasitic barnacles (*Sacculina*) the larva
attaches itself to a crab, throws
off its various appendages, and,
after other great degenerative
changes, enters its host. For

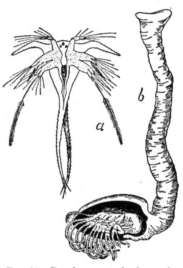

FIG. 62.—Development of a barnacle
(*Lepas*). *a*, larva ; *b*, adult.

a time, therefore, their development is toward greater complexity, but the later stages constitute a *retrograde metamorphosis.*

110. **More complex types.**—The larger, more useful, and
usually more familiar Crustacea belong to the second division (subclass Malacostraca). It comprises such animals as
the shrimps, crayfish, lobsters, crabs, and a number of other
forms which are at once distinguished from the preceding
by the constant number of segments composing the body.
Of these, five constitute the head, eight the thorax, and
seven the abdomen. The head segments are always fused
together, and with them one or more thoracic segments
unite to form a more or less complete cephalothorax. Also,

some of the head segments give rise to a great fold of the body-wall, the carapace, which extends backward and covers all or a part of the thorax, with which it may firmly unite, as in the crayfish. The appendages are usually highly specialized, and are made to perform a variety of functions.

111. The shrimps.—Among the simplest of these are the opossum-shrimps (Fig. 63) and their relatives, small trans-

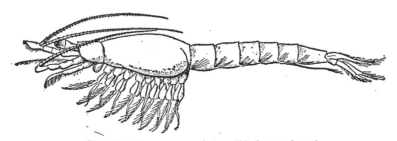

Fig. 63.—The opossum-shrimp (*Mysis americana*).

parent creatures often seen swimming in great numbers at the surface of the sea or hiding among the seaweeds along the shore. In general appearance they resemble crayfishes or prawns, but are readily distinguished by the two-branched thoracic feet. This "split-foot" character also occurs among many of the preceding Crustacea, and is generally a badge of low organization, tending to disappear in the more highly organized forms. In this and other respects the shrimps are especially interesting in their relation to the preceding Crustacea, and in the fact that they may closely resemble the ancestors of the modern prawns (Fig. 64), lobsters, crayfishes, and crabs.

112. Crayfishes and lobsters.—The last-mentioned species and their allies, usually large and familiar forms, constitute a group known as the decapods (meaning ten feet), referring to the number of thoracic feet. Among the members of this division probably none are more familiar than the crayfishes, which occur in most of the larger rivers and their tributaries throughout the United States and Europe. It is their habit to remain concealed in crevices of rocks

.or in the mouths of the burrows which they excavate, and
.from which they rush upon the small fish, the larvæ of

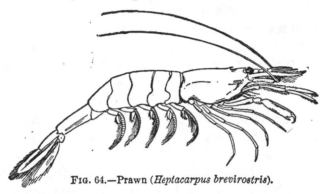

FIG. 64.—Prawn (*Heptacarpus brevirostris*).

many animals, and other equally defenseless creatures
which constitute their bill of fare. In turn they are
eagerly sought by certain birds and four-footed animals, and,
especially in France,
are extensively used for
food by man.

Closely related to
the crayfishes and dif-
fering but little from
them structurally are
the lobsters. In this
country they are con-
fined to the rocky coasts
from New Jersey to
Labrador, living upon
fish, fresh or otherwise,
various invertebrates,
and occasionally sea-
weeds. Far more than
the crayfish, the lobster
is in demand as an arti-
cle of food. By the aid
of nets or various traps

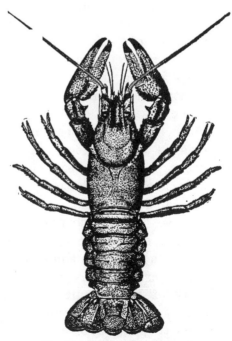

FIG. 65.—The crayfish (*Astacus*).

millions are caught each year, and to such an extent has their destruction proceeded that in many places they are well-nigh exterminated. At the present time, however, legislation, numerous hatcheries, and a careful study of their life habits is doing much to better matters and incidentally to put us in possession of many interesting zoological facts along this line, some of which will be mentioned later. Frequently the prawns, especially the larger ones, and a spiny lobster (*Palinurus*), are mistaken for crayfishes or lobsters, but they differ from them in the absence of the large grasping claws.

Along almost any coast some of these animals are to be found, often beautifully colored and harmonizing with the seaweeds among which they live, or so transparent that their internal organization may be distinctly seen. Farther out at sea other species swim in incredible numbers, feeding upon minute organisms, and in turn fed upon by numerous fishes and whales; and, especially on the Pacific coast, shrimp-fishing is an important industry.

113. The hermit-crabs.—The last of these long-tailed decapods is the interesting group of the hermit-crabs, which occur in various situations in the sea. In early life they take possession of the empty shell of some snail, and the protected abdomen becomes soft and flabby, while the appendages in this region almost completely disappear. The front part of the body, on the other hand, continually grows in firmness and strength, and is admirably adapted for the continual warfare which these forms wage among themselves. As growth proceeds the necessity arises for a larger shell, and the crab goes "house-hunting" among the empty shells along the shore, or it may forcibly extract the snail or other hermit from the home which strikes its fancy.

Many of the hermit-crabs enjoy immunity from the attacks of their belligerent relatives by allowing various hydroids to grow upon their homes. Others attach sea-anemones to their shells or to one of their large claws,

which they poke into the face of any intruder. While
the anemones or hydroids are made to do valiant service

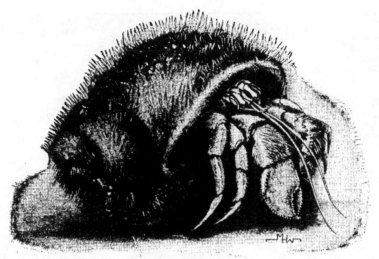

FIG. 66.—Hermit-crab (*Pagurus bernhardus*) in snail shell covered with *Hydractinia.*

with their nettle-cells, they also enjoy the advantages of
a large food-supply which is attendant upon the free ride.

114. **The crabs.**—The most highly developed Crustacea
are the crabs or short-tailed decapods which abound between
tide-marks alongshore, and in diminishing numbers extend
to great depths. The cephalothorax is usually relatively
wide, often wider than long, and the greatly reduced abdomen is folded against the under side of the thorax. Correlated with the small size of the abdomen, the appendages
of that region disappear more or less, but the remaining
appendages are similar to those of the crayfish or lobsters.
All these different parts, however, are variously modified in
each species to fit it for its own peculiar mode of life. In
some forms, such as the common cancer-crab (Fig. 67), the
legs are comparatively thick-set and possessed of great
strength, enabling them to defend themselves against most
enemies. On the other hand, there are the spider-crabs
with small bodies and relatively long legs, withal weak, and

yet so harmonizing with their surroundings that they **are**
as likely to survive as their stronger relatives.   In this

Fig. 67.—Kelp-crab (*Epialtus productus*) in upper part of figure ; to the right the
edible crab (*Cancer productus*), and the shore-crab (*Pugettia richii*).

connection it is interesting to note that the giant crab of
Japan, the largest crustacean, being upward of twenty feet
from tip to tip of the legs, is a spider-crab, constructed on

Fig. 68.—The fiddler-crab (*Gelasimus*).   Photograph by Miss Mary Rathbun

the same general pattern as our common coast forms.
Between these two extremes numberless variations exist,

some for known reasons, but more often not readily understood. And not only does the form vary, but the external
surface may be sculptured or beset with spines or tubercles
which frequently render the animal inconspicuous amid its
natural surroundings. Such an effect is heightened by the
presence of sponges, hydroids, and various seaweeds which
the crab often permits to gather upon its body.

115. **Pill-bugs and sandhoppers.**—Finally there remain the
groups of the pill- or sow-bugs (Isopods) and the sand-fleas
or sandhoppers (Amphipods). In the first of these the
body is usually small and compressed, the thorax more or
less plainly segmented, and the seven walking (thoracic)
legs are similar. In the female each leg bears at its base a
thin membranous plate which extends inward and hori-

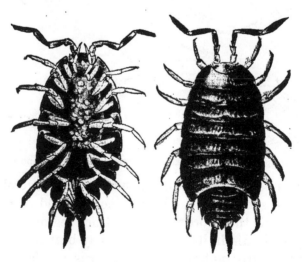

Fig. 69.—Isopod or pill-bug (*Porcellio laevis*).

zontally, thus forming on the under side of the body a
brood-pouch (Fig. 69) in which the young develop. As
one may readily discover in any of the common species,
the abdominal segments are more or less fused, and bear
appendages adapted for respiration and, in the aquatic
forms, for swimming.

The marine isopods occur in the sand, under rocks, and in the seaweeds; many are parasitic upon fishes; and the terrestrial forms (Fig. 69) are very common objects under old

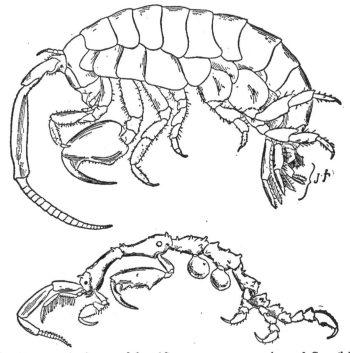

FIG. 70.—Amphipods or sand-fleas (*Gammarus*, upper species, and *Caprella*).

logs and in cellars, where they live chiefly on vegetable matter. In the sand-fleas the body is compressed from side to side, and while the thorax shows distinct segments, the legs are frequently dissimilar, and some may bear pincers. One of their most distinctive marks concerns the last three abdominal appendages, which are usually modified for leaping.

The sand-fleas (Fig. 70) are familiar objects to any one who has collected along the beach and has turned over the cast-up seaweeds, while numbers of small species often occur among the plants in our fresh-water ponds. Some most curious and highly modified forms, whose general appearance is shown in the lower part of Fig. 70, occur among hydroid colonies, with which their bodies harmonize in form and color. And, lastly, most bizarre creatures, known as "whale-lice," attach themselves to the skin of whales, of which each species acts as host for one or more kinds.

116. **Internal organization.**—Most Crustacea are carnivorous, preying upon almost any of the smaller animals within convenient reach; a much smaller number live on vegetable food; and there are many, such as the crayfishes, lobsters, and numerous crabs, which are also notorious scavengers. In these latter forms the food is held in one of the large pincers, torn into shreds by the other, and transferred to the mouth-parts, where, as in all Crustacea, it is soon reduced to a pulp by their rapid movements. In many species the food is now ready for the digestive process, but not so in the higher forms. If the stomach of any of these, for example, the crabs or crayfishes, be opened, three (Fig. 71, s) large teeth operated by powerful muscles will be noted, and beyond these a strainer consisting of many closely set hairs. In operation this "gastric mill" takes the food passed on from the mouth-parts, and crushes and tears it until fine enough to pass through the strainer, whereupon it is dissolved by the juices from the liver and is absorbed as it passes down the intestine.

The circulatory system is usually highly developed, and consists of a heart, in some species almost as long as the body, though usually shorter (Fig. 71), from which two or more arteries branch to all parts of the body. Here the blood, instead of emptying into definite veins, pours into a series of spaces or sinuses in among the muscles and other organs of the body, through which it makes its way back to the heart. During this return journey it is usually made to traverse definite respiratory organs, either situated upon the legs or, as feathery outgrowths, upon the sides of the body, and generally concealed under the carapace. A portion of the blood is also continually sent to the kidneys, which are located either at the base of the second antennæ (and known as green glands), as in the crayfishes or crabs, or on the second maxillæ (shell-glands) in many of the

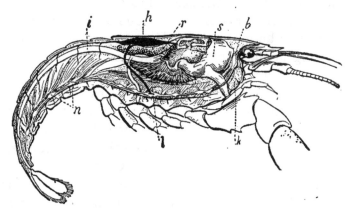

Fig. 71.—Dissection of crayfish. *b*, brain ; *h*, heart ; *i*, intestine ; *k*, kidney ; *l*, liver ; *n*, nerve-cord ; *r*, reproductive organ ; *s*, stomach, showing two teeth in position.

simpler crustaceans. Their method of operation is much like that of the kidneys in the earthworm.

117. **Nervous system and special senses.**—The nervous system also shows a decided resemblance to that of the annelids. The cerebral ganglia or brain is situated above the alimentary canal in the head, and connects with the ventrally lying cord by a collar. As in the earthworm, this ventral cord is double, and bears a pair of swellings or ganglia in each segment. In the crayfish, crabs, and other highly modified forms, where the segments tend to fuse, several of these ganglia may also unite, and except in early life their number cannot be determined.

Among the less specialized Crustacea the order of intelligence is low, though perhaps it may prove to be higher than is usually supposed when such forms have been more thoroughly studied. The following quotation relating to the lobster applies even more to the higher forms, the crabs : "Sluggish as it often appears when out of water and when partially exhausted, it is quite a different animal when free to move at will in its natural environment on the sea-bottom. It is very cautious and cunning, capturing its prey by stealth, and with weapons which it knows how to conceal. Lying hidden in a bunch of seaweed, in a crevice among the rocks, or in its burrow in the mud, it waits until its victim is within reach of its claws, before striking the fatal blow. The senses of sight and hearing are probably far from acute, but it possesses a keen sense of touch and of smell, and probably also a sense of taste."

Although enclosed in a horny and often very thick and strong armor, the sense of touch is very keen in the Crustacea and in arthropods generally. On many of the more exposed portions delicate hairs or pits connected with the nervous system occur in great abundance. Some of these, usually on the antennæ, undoubtedly serve in detecting odors, but the remainder are considered to be tactile. In the higher Crustacea, such as the crayfish, lobsters, and crabs, ears are usually found, consisting of sacs lined with similar delicate hairs, and containing several minute grains of sand, which in many cases make their way through the small external opening. Vibrations coming through the water gently shake the grains of sand, causing them to strike against the hairs which communicate with the nervous system—a very simple ear, yet sufficient for the needs of the animals.

The eyes of the Crustacea and arthropods in general are either simple or compound. The simple and frequently single eyes usually consist of a relatively few cells embedded in a quantity of pigment and connected with the nervous system. It is doubtful whether they perceive objects as anything more than highly blurred images, and perhaps they merely recognize the difference between light and darkness. The compound eyes, on the other hand, are remarkably complex structures, often borne on the tops of movable stalks, as in the common crabs and crayfishes. Each consists of an external transparent cornea, divided into numerous minute hexagonal areas corresponding to as many internal rods of cells, provided with an abundant nerve-supply. These latter elements may perhaps represent simple eyes grouped together to form the compound one; and it appears possible that each element may form a complete image of an object, as each of our eyes is known to do. On the other hand, many hold that the complete eye forms only one image, a mosaic, each element contributing its share.

118. **Growth and development.**—As we have seen, the simpler Crustacea hatch as minute larvæ (Fig. 62), and during their growth to the adult condition are especially subject to the attacks of multitudes of hungry enemies. In the higher forms, such as the crabs, some of these early transformations take place while the young are still within the egg and attached to the parent. Accordingly, the little ones are fairly similar to their parents, and their later history is very well exemplified by the lobster.

The eggs of the lobster are most frequently hatched in the summer months, usually July, after they have been carried by the parent for upward of a year. The young, about a third of an inch in length, at once disperse, undergo four or five moults during the next month, then, ceasing their swimming habits, settle to the bottom among the rocks. At this time, twice their original size, they closely resemble their parents, and their further development is largely an increase in size. " The growth of the lobster, and of every arthropod, apparently takes place, from infancy to old age, by a series of stages characterized by the growth of a new shell under the old, by the shedding of the outgrown old shell, a sudden increase in size, and the gradual hardening of the shell newly formed. Not only is the external skeleton cast off in the moult and the linings of the masticatory stomach, the esophagus, and intestine, but also the internal skeleton, which consists for the most part of a complicated linkwork of hard tendons to which muscles are attached."

119. **Peripatus (class Onychophora).**—It is generally believed that the Crustacea, insects, and spiders, together with their numerous relatives, trace their ancestry back to animals that bore a certain resemblance to the segmented worms. Most of these ancient types have long been extinct, but here and there throughout the earth we occasionally meet with them.

Among the most interesting of these are a few widely distributed species belonging to the genus *Peripatus* (Fig. 72), but as they are comparatively rare we shall dismiss them with a very brief description. They usually dwell in warm countries, under rocks and decaying wood, emerging at night to feed on insects, which they ensnare in the slime thrown out from the under surface of the head. Their external form, their excretory system, and various other organs are worm-like. On the other hand, the appendages are jointed, and one pair has been modified into jaws. The peculiar breathing organs characteristic of the insects are also present. *Peripatus* therefore gives us an interesting link between the worms and insects, and also affords an idea of the primitive insects from which the modern forms have descended.

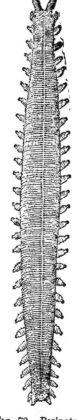

Fig. 72.—Peripatus (*Peripatus eiseni*). Twice the natural size.

120. **The centipeds and millipeds (class Myriapoda).**—Many of the myriapods—that is, the centipeds and thousand-legged worms—are familiar objects under logs and stones throughout the United States. The first of these (Fig. 73) are active, savage creatures, devouring numbers of small animals, which they sting by means of poison-spines on the tips of the first pair of legs. The bite of the larger tropical

species especially causes painful but not fatal wounds in man.

On the other hand, the millipeds (Fig. 74) or thousand-legs are cylindrical, slow-going animals, feeding on vegetable

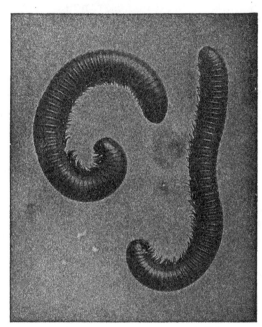

FIG. 73.—Centiped.
One-half natural size.

FIG. 74.—Thousand-legs or milliped (*Julus*).
Natural size.

substances without causing any particular damage, except in the case of certain species, which work injury to crops. When disturbed they make little effort to escape, but roll into a coil and emit an offensive-smelling fluid, which renders them unpalatable to their enemies.

All present a great resemblance to the segmented worms, as their popular names often testify; but, on the other hand, many points in their organization indicate a closer relationship to the insects. As in the latter, the head is distinct, and bears a pair of antennæ, the eyes, and two or three pairs of mouth-parts. The trunk is more worm-like, and consists of a number of similar segments, each bearing one or two pairs of jointed legs. In their internal organization the character of the various systems closely resembles that of the insects, and will be more conveniently described in that connection.

Among the myriapods the females are usually larger than the males. Some of the centipeds deposit a little mass of eggs in cavities in the earth and then abandon them, while others wrap their bodies about them and protect them until the young are hatched. The millipeds lay in the same situations, but usually plaster each egg over with a protective layer of mud. After several weeks the young appear, often like their parents in miniature, but in other species quite unlike, and requiring several molts to complete the resemblance.

CHAPTER X

121. Their numbers.—It has been estimated that upward of three hundred thousand named species of insects are known to the zoologist, and that these represent a fifth, or possibly a tenth, of those living throughout the world. Many of these species, as the may-flies and locusts, are represented by millions of individuals, which sometimes travel in such great swarms that they darken the sky. With nearly all of these the struggle for existence is fierce and unrelenting, and it is little wonder that such plastic animals have changed in past times and are now becoming modified in order to adapt themselves to new situations where food is more abundant and the conditions less severe. Owing to such modifications we find some species fitted for flying, others for running and leaping, or for a life underground, and many for a part or all of their lives are aquatic in their habits.

122. External features.—The body of an insect—the grasshopper, for example—consists of a number of rings arranged end to end, as we have seen them in the Crustacea and the segmented worms. In the abdomen these are clearly distinct, but in the thorax, and especially the head, they have become so intimately united that their number is a matter of uncertainty. These three regions—head, thorax, and abdomen—are usually clearly defined in most insects, but they are modified in innumerable ways in accordance with the animal's mode of life.

The head usually carries the eyes, a pair of feelers (antennæ), and three pairs of mouth-parts which may be fashioned into a long, slender tube to be used in sucking, and frequently as a piercing organ ; or they may be constructed for cutting and biting. The thorax bears three pairs of legs and usually two pairs of wings ; sometimes one pair or none. The appendages of the abdomen are usually small and few in number, or even absent.

123. Internal anatomy.—The restless activity of insects is proverbial. Some appear to be incessantly moving about either on the wing or afoot, and are endowed with comparatively great strength. Ants and beetles lift many times their own weight. Numerous insects are able to leap many times their own length, and others perform different kinds of work with a vigor and rapidity unsurpassed by any other class of animals. As is to be expected, the muscular system is well developed, and exhibits a surprising degree of complexity. Over five hundred muscles are required for the various movements of our own bodies, but in some of the insects more than seven times this number exist. The amount of food necessary to supply this relatively immense system with the required nourishment is correspondingly large. Many insects, especially in an immature or larval condition, devour several times their own weight each day. Their food may consist of the juices of animals or plants, which they suck out, or of the firmer tissues, which are bitten or gnawed off.

Not only do the mouth-parts stand in direct relation to the habits of the animal and to its food, but, as we have often noticed before, the internal organization is also adapted for the digestion and distribution of the nutritive substances in the most economical way. For this reason we find the alimentary canal differing widely in the various forms of insects. In each case it extends from the mouth to the opposite end of the animal, and ordinarily consists of a number of different parts. In the insect shown in

Fig. 75 the mouth soon leads into the esophagus, which in turn leads into the crop that serves to store up the food until ready for its entry into the stomach; or in some of the ants, bees, and wasps it may contain material which may be disgorged and fed to the young. In many cases the stomach is small and ill-defined as in Fig. 75, and again it may reach enormous dimensions, nearly filling the body. It may also bear numerous lobes or delicate hair-like processes, which afford a greater surface for the absorption of food. Behind the stomach are a number of slender outgrowths that are believed to act as kidneys. Beyond their insertion lies the intestine, which, like the stomach, is the subject of many modifications in the different kinds of insects.

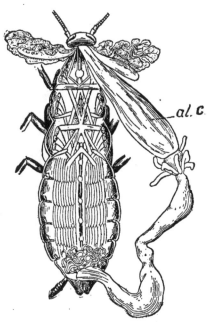

FIG. 75.—Cockroach, dissected to show alimentary canal, *al. c.*—After HATSCHEK and CORI.

The digested food is rapidly absorbed through the coats of the stomach and intestine and enters a circulatory system which reminds us of what exists in many of the Crustacea. The heart is situated above the digestive tract, and from it arteries pass out to different parts of the body. Here the blood leaves the vessels and is poured directly into the spaces among the viscera, whence it is finally conducted through irregular channels to the heart by its pulsations.

In the Crustacea the blood is made to pass through a respiratory system usually in the form of definite gills, and the oxygen with which it is charged is distributed to all parts of the body.   In the insects the blood serves almost entirely to carry the food, and the oxygen is conveyed through the animal by a remarkable contrivance found only in the insects, the spiders, and a few related forms.

124. **Respiratory system.**—If we examine an insect, the grasshopper for example, we find a number of small brown spots on each side of the abdomen, each of which under a magnifying-glass is seen to be perforated by a narrow slit. Carefully opening the body, we find that each slit is in communication with a white, glistening tube that rapidly branches and penetrates to all parts of the animal.   When the body is expanded the air rushes into the outer openings, on through the open tubes, and is distributed with great rapidity to all the tissues of the body.   In many insects some of these tubes connect with air-sacs which probably serve to buoy up the insect during its flights through the air.

125. **Wingless insects (Aptera).**—The simplest of all insects are the fishmoths and springtails, relatively small organisms covered with shining scales or hairs.   The first of these is occasionally seen running about in houses feeding upon cloth and other substances, while the latter live in damp places under stones and logs.   They are without wings, but are able to run rapidly and to leap considerable distances.   In addition to the ordinary appendages, the abdomen bears what are perhaps rudimentary legs, a fact which, together with their relatively simple structure, strengthens the belief that the insects have descended from centiped-like ancestors.

126. **Grasshoppers, crickets, katydids, etc. (Orthoptera).**— Rising higher in the scale of insect life, we arrive at the group of the cockroaches, crickets, grasshoppers, locusts, and other related insects.   Four wings are present, the first pair thickened and overlapping the second thinner pair.   The latter are folded lengthwise like a fan, which is said to have given the name *Orthoptera* (meaning straight-winged) to

this group of insects. These extend all over the world, being particularly abundant in the warmer countries, and their strong biting mouth-parts and voracious appetites render many of them dreaded pests to the farmer. The cockroaches are nocturnal in their habits, racing about at night, devouring victuals in the pantry and gnawing the bindings of books. During the day their flat bodies enable them to secrete themselves in crevices wherever there is sufficient moisture.

In the grasshoppers, locusts, katydids, and crickets the body is more cylindrical, and the hind pair of legs are often greatly lengthened for leaping. The crickets and katydids are nocturnal, the former remaining by day in burrows which they construct in the earth, the latter resting quietly in the trees. At night they feast upon vegetable matter principally, though

Fig. 76.—The Rocky Mountain locust.— After RILEY, from *The Insect World.*

some species are known to prey on small animals. Those insects we usually term grasshoppers (properly called locusts) are specially destructive to vegetation. Some species are strong fliers, and this, connected with their ability to multiply rapidly, renders them greatly dreaded pests. They have been described as flying in great swarms, forming black clouds, even hiding the sun as far as the eye could reach. The noise made by their wings resembled the roar of a torrent, and when they settled upon the earth every vestige of leaf and delicate twig soon disappeared.

The eggs of the majority of Orthoptera are laid in the ground, where they frequently remain through the winter. When hatched the young quite closely resemble the parents, and, after a relatively slight metamorphosis, assume the adult form.

127. **Dragon-flies, may-flies, white ants, etc.**—The dragon-, caddis-, may-flies, ant lions, and the white ants possess four

thin and membranous wings incapable of being folded.
These possess a network of delicate nervures, giving the
general name nerve-winged insects to these various small.
orders.   Of the forms mentioned above, all but the white
ants lay their eggs in the water, and the developing larvæ

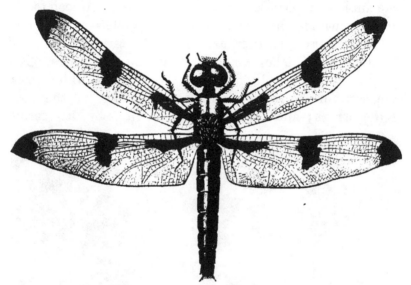

FIG. 77.—Dragon-fly (*Libellula pulchella*).

spend their lives in this medium until the time comes for their
complete metamorphosis into the adult.   The larvæ of the
caddis-flies protect themselves within a tube of stones or sticks
bound together with silken threads, which they usually
attach to the under side of stones in running water.   On
the other hand, the young of the dragon- and may-flies, provided with strong jaws, are active in the search of food and
very voracious.   In time they emerge from their larval skin
and the water in which they live, and after a life spent on
the wing they deposit their eggs and perish.   The adult
ant-lión, a type of the related order (Neuroptera), which
has somewhat the appearance of a small dragon-fly, lays its
eggs in light sandy soil.   In this the resulting larvæ excavate funnel-shaped pits, at the bottom of which they lie con-

cealed. Insects stumbling into their pitfalls are pelted with
sand, which the ant-lion throws at them with a jerky motion
of the head, and are speedily tumbled down the shifting
sides of the funnel to be seized and devoured.

While the white ants are not in any way related to the
true ants, they possess many similar habits. Associated in
great companies, they excavate winding galleries in old logs
and stumps, and, further, are most interesting because of
the division of labor among the various members. The
wingless forms are divided into the workers, which excavate, care for the young, and otherwise labor for the good
of the others; and into the soldiers, huge-headed forms,

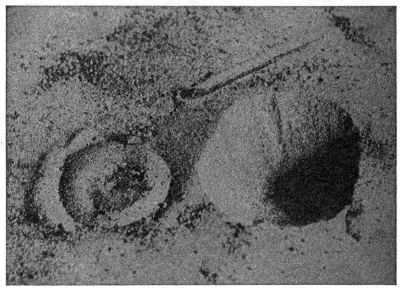

Fig. 78.—Ant-lion larva plowing its way through the sand (upper figure) while another is commencing the excavation of a funnel-shaped pit similar to one on right.
Photograph by A. L. Melander and C. T. Brues.

whose strong jaws serve to protect the colony. The remaining winged forms are the kings and queens. In the
spring many of the royalty fly away from home, shed their
wings, unite in pairs, and set about to organize a colony.
The queen rapidly commences to develop eggs, and in some

species her body becomes so enormously distended with these that she loses the power of locomotion and requires to be fed. A single queen has been known to lay eggs at the rate of sixty per minute (eighty thousand a day), and

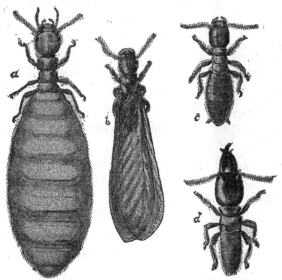

Fig. 79.—Termites or white ants. *a*, queen ; *b*, winged male ; *c*, worker ; *d*, soldier.

those destined to royal rank are so nursed that they advance farther in their development than the remaining sterile and wingless forms.

128. **The bugs (Hemiptera).**—The large and varied group of the bugs (*Hemiptera*) includes a number of semi-aquatic species, such as the water-boatmen, often seen rowing themselves along in the ponds by means of a pair of oar-shaped legs, in search of other insects. Somewhat similar at first sight are the back-swimmers, with like rowing habits, but unique in swimming back downward. Both of these bugs frequently float at the surface, and when about to undertake a subaquatic journey they may be seen to imprison a bubble of air to take along. Closely related are the giant water-bugs (Fig. 80), which often fly from pond to pond at night. In such flights they are frequently

attracted by lights, and have come to be called "electric-light bugs."

Among our most dreaded insect pests are the chinch-bugs—small black-and-white insects, but traveling in com-

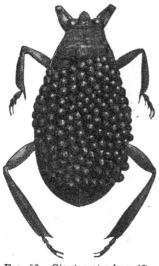

panics aggregating many millions. As they go they feed upon the stems and leaves of grain, which they devour with extraordinary rapidity. The squash-bug family is also extensive, and destructive to the young squash and pumpkin plants in the early spring.

The lice are small, curiously shaped bugs, which suck the blood of other animals. The plant-lice, also small, suck the juices of plants, and are often exceedingly destructive. This is especially true of the phylloxera, a plant-louse which causes annually the loss of millions of dollars among the vine-yards of this and other countries.

Fig. 80.—Giant water-bug (*Serphus dilatatus*), with eggs attached.

Even more destructive are the scale-insects, curiously modified forms, of which the wingless females may be found on almost any fruit-tree and on the plants in conservatories, their bodies covered with a downy, waxy, or other kind of covering, beneath which they remain and lay their eggs.

129. **The flies (Diptera).**—The group of the Diptera (meaning two-winged) includes the gnats, mosquitoes, fleas, house-flies, horse-flies (Fig. 81), and a vast company of related forms. Only a single pair of wings is present, the second pair being rudimentary or fashioned into short, thread-like appendages known as balancers, though they probably act as sensory organs and are not directly concerned with flight. The mouth-parts are adapted for piercing and sucking. The eyes, constructed on the same plan

as those of the Crustacea, are comparatively large, and are frequently composed of a great number of simple eyes united together, upward of four thousand forming the eye of the common house-fly.

These insects are widely distributed throughout the world, where they inhabit woods, fields, or houses as best suits their needs. Their food is varied. Some suck the juices of plants, others attack animals, and, while many are troublesome pests, others, especially in the early stages of their existence, are of great benefit.

FIG. 81.—Horse-fly (*Therio-plectes*).

130. **Typical forms.**—Owing to the widely different habits and structure of the members of this group, we shall briefly consider two examples, the mosquito and the house-fly, which will give us a fairly good idea of the characteristics of all. The eggs of the mosquito are laid in sooty-looking masses on the surface of stagnant pools. Within a very short time the young hatch, and, owing to their peculiar swimming movements, are known as "wrigglers." They are then active scavengers, devouring vast quantities of noxious substances and performing a valued service. They frequently rise to the surface, take air into the tracheal system, which opens at the posterior end of the body, and descend again. After an increase in growth and many internal changes resulting in a chrysalis-like stage, they rise to the surface, split the shell, and, using the latter as a float, carefully balance themselves and soon fly away.

The house-fly usually lays its eggs in decaying vegetable matter, and the young, maggot-like in form, are active scavengers. They too undergo deep-seated changes during the next few days, finally transforming into the adult.

Many of this great group of the flies spend their early life in the water or other medium acting as scavengers; but, on the other hand, numbers attack domestic and other animals, and throughout their entire lives are an intolerable plague.

131. **The beetles (Coleoptera).**—Owing to the ease of preservation and their bright colors, the beetles have probably been more widely collected than other insects. Fully ten

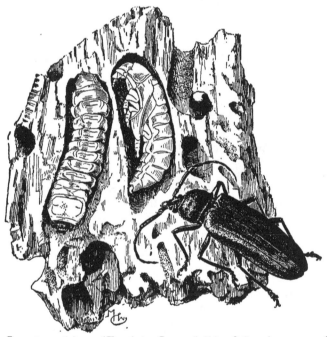

Fig. 82.—Long-horned borer (*Ergates*). Larva (left-hand figure), pupa, and adult insect.

thousand distinct species are known in North America alone. They are all readily recognized by the two firm, horny sheaths enclosing the two membranous wings, which alone are organs of flight. The mouth is provided with jaws, which are used in gnawing. Some prey on noxious insects or upon decaying vegetable or animal matter, and are often highly beneficial; but others attack our trees and domestic animals, and work incalculable damage.

In some of the stag- or wood-beetles (Fig. 82), which
we may select as types, the adults are often found crawling
about on or beneath the bark of trees, living on sap or
small animals. The eggs laid in these situations develop
into grub-like larvæ, which bore their way through living
or dead wood, and in this condition sometimes live four or
five years. They then transform into quiescent pupæ (Fig.
82), which finally burst their shells and emerge in the
adult form. Others, like water-beetles and the whirligig-
beetles, whose mazy motions are often seen on the surface
of quiet streams, pass the larval period in the water.
Under somewhat different conditions we find the potato-
bugs, lady-bugs, fire-flies, and their innumerable relatives,
but the changes they undergo in becoming adult are essentially the same as those described for the other members of
the order.

132. **The moths and butterflies (Lepidoptera).**—The moths
and butterflies occur all over the world. In their mature

Fig. 83.—Monarch-butterfly (*Anosia plexippus*). From photograph by A. L. MELAN-
DER and C. T. BRUES.

state they are possessed of a grace of form and movement
and a brilliancy of coloration that elicit our highest admi-
ration. The mouth-parts are developed into a long proboscis, which may be unrolled and used to suck the nectar
out of flowers, though in many of the adult moths, which
never feed, it may remain unused. The wings, four in
number, are covered with beautiful overlapping scales that

adhere to our fingers when handled. This feature, and the general plan of the body, which is much the same throughout the group, enables us to recognize most of them at once.

Fig. 84.—The silver-spot (*Argynnis cybele*). Photograph by A. L. Melander and C. T. Brues.

133. The ants, bees, wasps, etc. (Hymenoptera).—The ants, bees, and wasps are the best-known insects belonging to this order. They are characterized by four membranous wings, by biting and sucking mouth-parts, and the female is often provided with a sting. All undergo a complete metamorphosis. The eggs may be laid in the bodies of other insects, or they may be placed in marvelously constructed homes, and be the objects of the greatest attention, the parents or attendants often risking or losing their lives in their defense. The members of this order have long attracted attention, largely on account of their remarkable instinctive powers. They live in highly organized communities, and certain of their characteristics may be illustrated by a study of some of the more familiar forms.

CHAPTER XI

134. General characters.—In this group, comprising the spiders, mites, and a large assemblage of related species, we again meet with great differences in form and structure which fit them for lives under widely different conditions. The three regions of the body, head, thorax, and abdomen, so clearly marked in the insects, are here less plainly defined. The head and thorax are usually closely united, and in the mites the boundaries of the abdomen are also indistinct. The appendages of the head are two in number, and probably correspond to the antennæ and mandibles of other Arthropods. In the scorpions and some species of mites these are furnished with pincers for holding the prey, and in other forms they act as piercing organs. Usually the thorax bears four pairs of legs, a characteristic which readily separates such animals from the insects.

The internal organization differs almost as much as does the external. In many species it shows a considerable resemblance to that of some insects, but in others, especially those of parasitic habits, it departs widely from such a type. Respiration is affected by means of tracheæ, or lung-books, which consist of sacs containing many blood-filled, leaf-like plates placed together like the leaves of a book.

Usually, as in the insects, the young hatch from eggs which are laid, but in the scorpions and some of the mites the young develop within the body and at birth resemble the parent. Almost all of these organisms live either as

143

parasites or as active predaceous animals upon other animals. For this purpose many are provided with keen senses for detecting their prey and poisonous spines for despatching it.

135. **The scorpions.**—Owing to the stout investing armor, the strong pincers, and the general form of the body, the scorpions might at first sight be mistaken for near relatives of the crayfish or lobster. A more careful examination will show that the two pairs of pincers probably correspond to the antennæ and mandibles of the Crustacea that have become modified for seizing the food. The swollen part of the animal lying behind the four pairs of legs is a part of the abdomen, of which the slender " tail " constitutes the remainder. On the tip of the tail is a curved spine supplied with poison glands. Several pairs of eyes are borne on the dorsal surface of the head and thorax, while

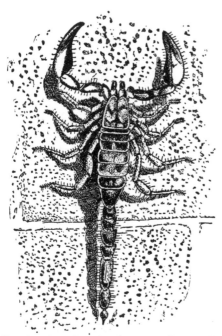

Fig. 85.—Scorpion, showing pincer-like mouth-parts and spine-tipped tail.

on the under side of the animal several slit-like openings lead into as many small cavities containing the lung-books.

The scorpions are the inhabitants of warm countries, where they may be found under sticks and stones throughout the day. At night they leave their homes in search of food, which consists chiefly of insects and spiders. These are seized by means of .the pincers, and the sting is driven into them with speedily fatal results. It is doubtful if the poison causes death in man, but the sting of some of the larger species, which measure five or six inches in length, may produce certain disorders chiefly affecting the circulation. In this country there are upward of thirty species, most of which are comparatively small.

136. **The harvestmen.**—The harvestmen or daddy-long-legs are small-bodied, long-legged creatures which resemble in general appearance several of the spiders. They differ from them, however, in the possession of claws corresponding to the smaller ones of the scorpion, and in their method of respiration, which is similar to that of insects. During the day they conceal themselves in dark crevices or stride slowly about in shaded places; but at night they emerge into more open districts and capture small insects, from which they suck the juices.

137. **The spiders.**—The spiders are world-wide in their distribution, and are a highly interesting group, owing chiefly to their peculiar habits. Examining any of our familiar species, it will be seen that the united head and thorax are separated by a narrow stalk from the usually distended abdomen. To the under side of the former are attached four pairs of long legs, a pair of feelers, and the powerful jaws supplied with poison-sacs, while eight shining eyes are borne on the top of the head. On the abdomen, behind the last pair of legs, are small openings into the lung cavities which contain a number of vascular, leaf-like projections known as lung-books. In some species a well-marked system of tracheæ are also present. At the hinder end of the body are four or six little projections, the spinnerets, each of which is perforated with many holes. Through these the secretion from the glands beneath is squeezed out in the form of excessively delicate threads, often several hundred in number, which harden on exposure to the air. According to the use for which these are intended, they may remain a tangled mass or become united into one firm thread; and according to the habits of the animal, they may be used for enclosing their eggs,

for lining their burrows, or for the construction of webs of the most diverse patterns.

138. **The habits of spiders.**—Many species of spiders, some of which are familiar objects in fields and houses, construct sheets of cobweb with a tube at one side in which they may

Fig. 86.—A tarantula-spider (*Eurypelma lentzii*). Natural size. Photograph by A. L. Melander and C. T. Brues.

lie in wait for their prey or through which they may escape in times of danger. In the webs of the common orb- or wheel-weavers several radial lines are first constructed, and upon these the female spider spins a spiral web. Resting in the center of this or at the margin, with her foot on some of the radial threads, she is able to detect the slightest tremor and at once to rush upon the entangled captive.

Some of the bird-spiders and their allies, living in tropical America, and attaining a length of two inches, construct web-lined burrows in the ground. From these they stalk their prey, which consists of various insects and even

small birds. These are almost instantly killed by the poison-
fangs, and are then carried to the burrow, where the juices
of the body are extracted.

The trap-door spiders of the southwestern section of the
United States also dig tunnels, which they cover with a
closely fitting lid com-
posed of earth. Raising
this they come out in
search of insects, but if
sought in turn, they dash
into the burrow, closing
the door after them, and
holding it with such firmness that it is rarely forced
open. If this should happen, there are sometimes
blind passage-ways, also
closed with trap-doors,
which usually baffle the
pursuer.

Finally, there are
among the thousand spe-

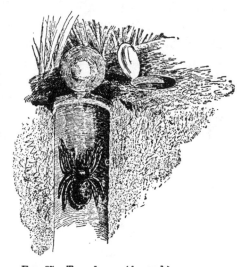

Fig. 87.—Trap-door spider and burrow
(*Cteniza*).

cies of spiders in the United States a considerable proportion which construct no definite web. Many of these may
be seen darting about in the sunshine on old logs and
fences, often trailing after them a thread which may support them if they fall in their active leaping after insects.

139. **Breeding habits.**—The male spiders are usually much
smaller than the females, and some species are only one-
fifteenth as long as the female and one one-hundredth of
its weight. They are usually more brilliantly colored, more
active in their movements, yet rarely spinning their own
webs and capturing their own food, preferring to live at
the expense of the female. At the breeding season the
males of several species make a most interesting display

of their colors, activity, and gracefulness before the females; and the latter, after watching these exhibitions, are said to select the one who has "shown off" in the most pleasing fashion. The life after this may be stormy, resulting in the death of the male; but ordinarily the results are not so disastrous, and in a little while the female deposits her eggs in cases which she spins. In these the young develop, sometimes wintering here, and emerging in the spring to scamper about in search of food, or to drift through the air to more favorable spots on fluffy masses of cobweb.

Few groups of animals are more interesting objects of study and more accessible. Their bites are rarely more serious than those of the mosquito—never fatal; and a

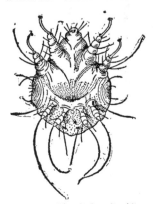

careful study of any species, however common, will undoubtedly bring to light many interesting and unknown facts.

140. **The mites and ticks.**—The mites and ticks are the simplest and among the smallest of the animals belonging to this group. To the attentive observer they are rather common objects, with homes in very different situations. Some occur on living and decaying vegetation, in old flour and unrefined sugar, while oth-

FIG. 88.—The itch-mite (*Sarcoptes scabei*).

ers live in fresh water and a few in the sea. Almost all tend toward parasitism. Some of the insects which they pierce and destroy are a pest to man, but on the other hand some are intolerable owing to the diseases they produce.

As to other parasitic organisms, degradation of structure is manifest. The respiratory system, so important to the active life of the insects, may be absent, the animal breathing through its skin. The circulatory system may be wanting, the blood occupying spaces among the various organs being swept about by the animal's movements. And many other peculiarities have arisen which fit them for their different modes of life.

141. **The king crab (Limulus).**—The king crab may be found crawling over the bottom or plowing its way through the sand and mud in many of the quiet bays from Maine to Florida. The large head and thorax of these animals are united into a horse-shoe-shaped piece, behind which lies the triangular abdomen. On the curved front surface of the former are a pair of small median eyes, and farther outward are two larger compound ones. On the ventral side are six pairs of appendages, instrumental in capturing and tearing the small animals that serve as food, and functioning in connection with the terminal spine as locomotor organs. On the

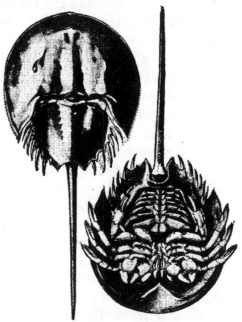

Fig. 89.—The king or horseshoe crab (*Limulus polyphemus*).

ventral surface of the abdomen are numerous plate-like flaps which serve in respiration, and in the imperfect swimming movements in which these animals occasionally indulge.

These relatively large and clumsy creatures are the remnant of a great number of strange, uncouth animals that inhabited the earth in past ages. Many of them show a close resemblance to the scorpions. The anatomy and development also show certain points of resemblance, and by some are thought to give us an idea of the ancient type of spider-like animal from which the modern forms have descended.

CHAPTER XII

142. **General characters.**—The division of the echino-derms includes the starfishes, sea-urchins, serpent- or brittle-stars, sea-cucumbers, and crinoids or sea-lilies. All are ma-rine forms, and constitute a conspicuous portion of the animals along almost any coast the world over. From these shallow-water situations they extend to the greatest depths of the ocean, and the bodily form possesses a great number of variations, adapting them to lives under such diverse conditions; and yet there is perhaps no group of organisms so clearly defined or exhibiting so close a resem-blance throughout. At one time it was thought that their radial symmetry was an indication of a close relationship to the cœlenterates, but more careful study has shown them to be much more highly developed than this latter group, and widely separated from it. A skeleton is almost always present, consisting of a number of calcareous plates embed-ded in the body-wall, and often supporting numbers of pro-tective spines, which fact has given to the group the name Echinoderm, meaning hedgehog skin.

143. **External features.**—The body of a starfish (Fig. 90) consists of a more or less clearly defined disk, from which the arms, usually five in number, radiate like the spokes of a wheel. At the center of the under side the mouth is located, and from it a deep groove, filled with a mass of tubular feet, extends to the tip of each arm. Innumerable calcareous plates firmly embedded in the body-wall serve.

150

for the protection of the internal organs, and at the same time admit of considerable movement.

In the brittle-stars (Fig. 91) the central disk is much more sharply defined than in the preceding forms, and the long snake-like arms are capable of a very great freedom of movement, enabling the animal to glide over the sea-bottom, or through the crevices of the rocks, at a surprising rate.

In several species, otherwise closely resembling those

FIG. 90.—Starfish (*Asterias ocracea*), ventral view. One-half natural size.

in Fig. 91, the arms divide repeatedly. These are the so-called basket-stars, living in the deeper waters of the sea, where they, like other brittle-stars, act as scavengers and devour large quantities of decomposing plant or animal remains.

At first sight the globular spiny sea-urchins (Fig. 93) would scarcely be recognized as close relatives of the star-fishes. A closer examination, however, shows the mouth to be located on the under side of the body; from it five rows of feet radiate and terminate close to the center of the dorsal side, and the arrangement of the plates forming the

skeleton indicate that the sea-urchin is comparable to a
starfish, with its dorsal surface reduced to insignificant
proportions.

In the sea-urchins the calcareous plates possess a great
regularity, and are so closely interlocked that they prevent

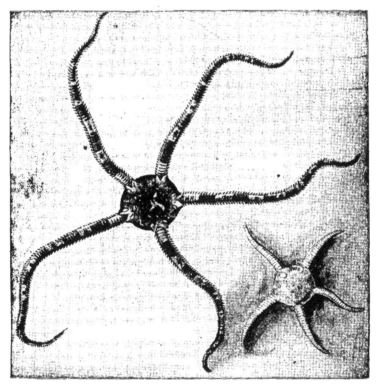

FIG. 91.—Brittle- or serpent-stars (species undetermined). Natural size.

any motion of the body-wall. Also, each plate is usually
provided with highly developed spines, movable upon a ball-
and-socket joint. These spines serve for locomotion, and,
in some instances, for conveying food to the mouth. A
considerable number of sea-urchins show an irregularity in
form which destroys to a corresponding degree the radial
symmetry. This is due to various causes, but especially to
a compression of the body, which, in the "sand-dollars,"

has resulted in the production of a thin, cake-like form (Fig. 94).

If the spherical body of a sea-urchin were to be stretched in the direction of a line joining the mouth and the center

Fig. 92.—Basket-star (*Astrophyton*). One-half natural size.

of the dorsal surface, a form resembling a sea-cucumber (Fig. 95) would be the result.   These latter organisms live among crevices of the rocks, embedded in the mud or burrowing in the sand at the bottom of the sea.   In such situations they are well protected, and a highly developed skeleton, such as that of the sea-urchin, would not only be of little value, but a positive hindrance to locomotion.   The skeleton, therefore, is much reduced, consisting of a few scattered calcareous plates embedded in the fleshy bodywall.   Another peculiar feature is almost universally present, in the form of a circlet of tentacles surrounding the mouth, which serve either for the purpose of respiration, for locomotion, or to convey food to the mouth.

A very good imitation of the general plan of a sea-lily or crinoid (Fig. 96) could be made by attaching a serpent-star, especially one of the basket-stars, by its dorsal side to a stalk. In the crinoids the numerous branches of the arms are comparatively short, and in the arrangement of the internal organs there are numerous differences, but for all that the resemblance of these organisms to the other echinoderms is undoubted.

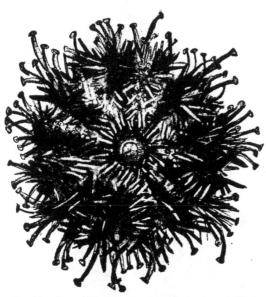

Fig. 93.—Sea-urchin (*Strongylocentrotus purpuratus*). Natural size.

144. Haunts. — The greater number of starfishes occur alongshore, slowly crawling about in search of food, or concealed in dark crevices of the rocks, where they may often be found as the tide goes out, and we know that in gradually lessening numbers other species lead similar lives at different levels far down in the dark and gloomy depths. In these same locations the sea-urchins occur, sometimes singly, but more usually associated in great numbers, several species excavating hollows in the rocks, within which they obtain protection. The brittle-stars and sea-cucumbers may also be found occasionally in open view, but more often they make their way about in search of food buried in the sand. The crinoids are usually inhabitants of deeper water, where they are found associated often in great numbers. A few species upon attaining the adult condition separate from the stalk, and are able to move about (Fig. 97), but the remaining species never shift their position.

. **145. The organs of defense and repair of injury.**—As we have seen, the body-wall of the echinoderms is provided with a series of plates, often bearing spines which serve as organs of defense, and to protect the internal organs. The starfishes and sea-urchins also possess numerous modified spines (*pedicellaria*) scattered over the surface of the body, which have the form of miniature birds' beaks, fastened to slender muscular threads. During life these jaws continually open and close, and it is said they clean the body of *débris* that settles on it; but on the other hand there are several reasons for the belief that they also act as organs of defense. Thus protected, the natural enemies of echinoderms appear to be relatively few, and are confined chiefly to some of the fishes whose teeth are especially modified for crushing them. In this way, and owing to the action of the breakers, they suffer frequent injury, but many species exhibit to a remarkable degree the ability to regenerate lost parts. Experiments show that if all the arms of a starfish be separated from the disk the latter will within two or three months renew the arms; and a single arm with a part of the disk is able to renew the missing portions in about the same length of time.

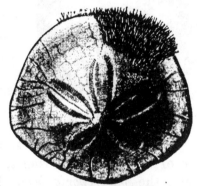

Fig. 94.—Sand-dollar, a flat sea-urchin. Natural size.

The brittle-stars, as their name indicates, are usually excessively delicate, often dropping all of their arms upon the slightest provocation; but here again the ability is present to develop the lost portions.

Sea-cucumbers resent rough treatment by vigorously contracting their muscular walls and removing from the body almost the entire digestive tract, the respiratory tree, and a portion of the locomotor system; but some species, at least, renew them again. In some of the starfishes and

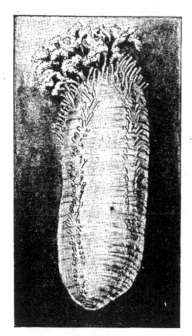

FIG. 95.—Sea-cucumber (*Cucumaria* sp.). Natural size.

brittle-stars portions of the body appear to be voluntarily detached and to develop into new individuals, and it is thought that such self-mutilation is a normal method of reproduction.

146. Locomotor system.—One of the most characteristic and remarkable features of the echinoderms is the water-vascular system, a series of vessels containing water which serve in the process of locomotion. Their arrangement and mode of operation are, with slight modifications, the same throughout the group, and may be readily understood from their study in the starfish.

On the dorsal surface of a starfish, in the angle between two of the arms, is a round, slightly elevated, calcareous plate, the *madreporic body* (Fig. 98, *m.p.*), which under the microscope appears full of holes, like the "rose" of a watering-pot. This connects with a tube that passes to the opposite side of the body, where it enters a canal completely encircling the mouth. On this ring-canal a number of sac-like reservoirs with muscular walls are attached, and from it a vessel extends along the under surface of each arm from base to tip. Each of these radial water-mains gives off numerous lateral branches that open out into small reservoirs similar to those located on the ring-canal, and a short distance beyond communicate through the wall of the body with one of the numerous

tube-feet, which, as we have seen, are slender tubular organs, many in number, filling the grooves on the ventral
surface of each arm. This entire system of tubes and
reservoirs is full of water, taken in, it is said, through the
perforated plate, and, when the starfish wishes to advance,
many of the little reservoirs con-
tract, forcing water into the cavity of the feet, with which they
are in communication, thus extending the extremity of the tubes
a considerable distance. The
terminal sucker of each foot, acting upon the same principle as
those on the cuttlefish, attaches
firmly to some foreign object,
whereupon the muscles of the
foot contract, drawing the body
toward the point of attachment.
This latter movement is similar
to that of a boatman pulling himself to land by means of a rope
fastened to the shore. When the
shortening of the tube-feet has
ceased, the sucking disks release
their attachment, project themselves again, and this process is
repeated over and over. At all
times some of the feet are contracting, and a steady advance of
the body is the result.

FIG. 96.—Sea-lily or crinoid.

This method of locomotion
also obtains in the sea-urchins and cucumbers, but in the
serpent-stars the tube-feet have become modified into feelers, and the animal moves, often rapidly, by means of twisting movements of the arms. The feet have this character
also in the crinoids, where the animal is generally without

the power of locomotion. In some of the sea-cucumbers five equidistant rows of tube-feet extend from one end of the body to the other, and the animal crawls worm-like upon any side that happens to be down; but certain species living in the sand, where tube-feet will not work satisfactorily, have lost all traces of them, and creep like an earthworm from place to place. In all the sea-cucumbers the feet, situated near the mouth, have been curiously modified to form a circlet of tentacles, which range in form from highly branched to short and thick structures, and in function from respiratory organs and those of touch to contrivances for scooping up sand and conveying it to the mouth.

Fig. 97.—An unattached crinoid (*Antedon*). One-half natural size.

147. Food and digestive system.—In the echinoderms the body-wall is comparatively thin (Fig. 98), and encloses a great space, the body-cavity, in which the digestive and reproductive organs are contained. As the former in various species is adapted for acting upon very different kinds of food, it shows many modifications; but there are a few principal types which may be briefly considered.

In the starfishes the mouth enters almost directly into the cardiac division of the stomach, a capacious, thin-walled sac, much folded and packed away in the disk and bases of the arms (Fig. 98, *b*). This in turn leads into the second pyloric portion (*a*), with thicker walls and dorsal, to the first, from which a short intestine leads to the exterior, near the center of the disk. Another conspicuous and important feature is the so-called liver, consisting of a pair of closely branched, fluffy glands (*l*), extending the entire length of each arm and opening into the pyloric stomach.

The starfishes are carnivorous and highly voracious, devouring large numbers of barnacles and mollusks which happen in their path. If these are small and free they are taken directly into the stomach, but when one of relatively large size is encountered the starfish settles down upon it, and, slowly pushing the cardiac stomach through the mouth, envelops it in the folds. Digestive fluids are now poured over it, and the victim is speedily despatched and in a partly digested condition is gradually absorbed into the body, leav-

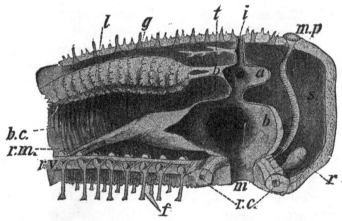

Fig. 98.—Dissection of starfish to show: *a*, pyloric stomach; *b*, bile-ducts (above), cardiac stomach (below); *b.c.*, body-cavity; *f*, feet; *g*, spines; *i*, intestine; *l*, liver; *m*, mouth; *m.p.*, madreporic plate; *r*, reservoir; *r.c.*, ring canal; *r.m.*, stomach retractor muscle; *r.v.*, radial vessel; *s*, stone canal; *t*, respiratory tree.

ing the shell and other indigestible matters upon the exterior. Oysters and clams close their shells when thus attacked, but a steady, continuous pull on the part of the starfish finally opens them, and the stomach is spread over the fleshy portions with speedily fatal results. In the interior of the body the food is transferred to the pyloric stomach, subjected to the action of the liver, and when completely dissolved is borne to all parts of the body.

The digestive system of the starfishes, with its various subdivisions and appendages, is in some respects more complicated than in the other classes. This is most strikingly the case with the serpent-stars, where the entire system for disposing of the minute animals and plants on which it feeds consists of a simple sac communicating with the exterior by a single opening—the mouth.

In the sea-cucumbers large quantities of sand are taken into the body, and the minute organisms and organic matter are digested from it. In the sea-urchins the mouth is provided with five teeth, and the food consists of minute bits of seaweeds, which these snip off. Such diets evidently require a comparatively simple digestive apparatus, for in both it consists throughout its whole extent of a tube of equal caliber, in which the various divisions of esophagus, stomach, and intestine are little, if at all, defined. This is usually somewhat longer than the body, and therefore thrown into several loops; and in the sea-cucumbers its last division is expanded and furnished with more highly muscular walls, which aid in respiration.

148. **Development.**—With but a few exceptions, the eggs of the echinoderms are laid directly in the surrounding water, and for many days the exceedingly minute young are borne great distances in the tidal currents. During this period they show no resemblance to their parents, and only after undergoing remarkable transformations do they assume their permanent features. In every case they have a five-rayed form in early youth, but in several species of starfishes additional arms develop until there may be as many as twenty or thirty.

CHAPTER XIII

THE CHORDATES

149. **General characters.**—Up to the present time we have been studying the representatives of a vast assemblage of animals whose skeletons, if they have any at all, are located on the outside of the body. In the corals, the mighty company of arthropods, and the echinoderms, it is external. On the other hand, we shall find that the animals we are now about to consider, the fishes, frogs, lizards, birds, and mammals, are in possession of an internal skeleton. In some of the simpler fishes and in a number of more lowly forms (Fig. 99) it is exceedingly simple, and consists merely of a gristle-like rod, the *notochord* (Fig. 101, *nc*), extending the length of the body and serving to support the nervous system, which is always dorsal. This is also the type of skeleton found in the young of the remaining higher animals, but as they grow older the notochord gives way to a more highly developed cartilaginous or bony, jointed skeleton, the vertebral column.

In the young of all these back-boned or chordate animals, the sides of the throat are invariably perforated to form a number of gill-slits. In the lower forms these persist and serve as respiratory organs, but in the higher animals they disappear in the adult. The chordates are thus seen to be distinguished by the possession of a dorsal nervous cord supported by an internal skeleton and by the presence of gill-slits, characters which separate them widely from all invertebrates.

The chordates may be divided into ten classes, seven of

which (the lancelets, lampreys, fishes, amphibians, reptiles, birds, and mammals) are true vertebrates, while the others embrace several peculiar animals of much simpler organization.

150. **The ascidians.**—Among the latter are a number of remarkable species belonging to the class of ascidians or sea-squirts (Fig. 99).

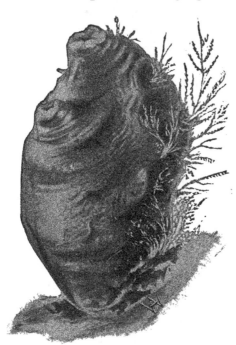

These are abundantly represented along our coasts, and are readily distinguished by their sac-like bodies, which are often attached at one end to shells or rocks. On the opposite extremity two openings exist, through which a constant stream of water passes, bearing minute organisms serving as food. When disturbed they frequently expel the water from these pores with considerable force, whence the name "sea-squirt." While many lead solitary lives, numerous individuals of other species are often closely packed together in a jelly-like pad attached to the rocks, and others not distantly related are fitted to float on the surface of the sea.

Fig. 99.—Ascidian or sea-squirt.

The young when hatched resemble small tadpoles both in their shape and in the arrangement of some of the more important systems of organs. For a few hours each swims about, then selecting a suitable spot settles down and adheres for life. From this point on degeneration ensues.

The tail disappears, and with it the notochord and the greater part of the nervous system. The sense-organs vanish, the pharynx becomes remodeled, and numerous other changes occur, leaving the animal in its adult condition, with little in its motionless, sac-like body to remind one of a vertebrate.

151. **The vertebrates.**—Since the remainder of this volume is concerned with the vertebrates it will be well at the outset to gain some knowledge of their more important characteristics. One of the most apparent is the presence of a jointed vertebral column, composed of cartilage or bone, which supports the nervous system. To it are also usually attached several pairs of ribs, two pairs of limbs, either fins, legs, or wings, and in front it terminates in a more or less highly developed skull. In the space partially enclosed by the ribs, the body-cavity, a digestive system is located, which consists of the stomach and intestine, together with the attached liver and pancreas. The circulatory system is also highly organized, and consists of a muscular heart, arteries, and veins which ramify throughout the body. Breathing, in the aquatic animals, is carried on by means of gills, and in the air-breathing forms by means of lungs, which, like the gills, effect the removal of carbonic-acid gas and the absorption of oxygen. The nervous system, consisting of the brain situated in the head and the spinal cord extending through the body above the back-bone, even in the lower vertebrates, is far more complex than in the invertebrates. The sense-organs also attain to a high degree of acuteness, and in connection with the highly organized nervous system enable these forms to lead far more varied and complex lives than in any of the animals heretofore considered.

CHAPTER XIV

THE FISHES

152. General characters.—In a general way the name fish is applied to all vertebrates which spend the whole of their life in the water, which undergo no retrograde metamorphosis, and which do not develop fingers or toes. Of other aquatic chordates or vertebrates the ascidians undergo a retrograde metamorphosis, losing the notochord, and with it all semblance of fish-like form. The amphibians, on the other hand, develop jointed limbs with fingers and toes, instead of paired fins with fin rays. A further comparison of the animals called fishes reveals very great differences among them—differences of such extent that they cannot be placed in a single class. At least three great groups or classes must be recognized: the Lancelets, the Lampreys, and the True Fishes. The general characters of all these groups will be better understood after the study of some typical fish, that is one possessing as many fish-like features as possible, unmodified by peculiar habits. Such an example is found in the bass, trout, or perch. In either fish the pointed head is united, without any external sign of a neck, to the smooth, spindle-shaped body, which is thus fitted for easy and rapid cleaving of the water when propelled by the waving of the powerful tail (Fig. 100). A keel also has been provided, enabling the fish to steer true to its course. This consists of folds of skin arising along the middle line of the body, supported by numerous bony spines or cartilaginous

164

rays.   These are the unpaired fins, as distinguished from
the paired ones, which correspond to the limbs of the higher
vertebrates.   In the bass or perch the latter are of much
service in swimming, and are also most important organs in
directing the course of the fish upward or downward, or for

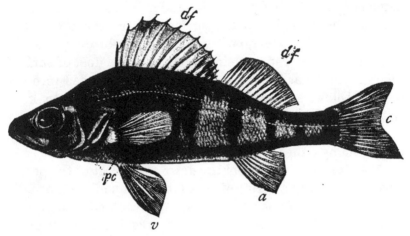

FIG. 100.—Yellow perch (*Perca flavescens*).   *df*, dorsal fins ; *pc*, pectoral fin ; *v*, ventral fin ; *a*, anal fin ; *c*, caudal fin.

aiding the tail in changing the course from side to side;
or they may be used to support the animal as it rests upon
the bottom in wait for food; and, finally, they may serve to
keep the body suspended at a definite point.

In addition to an internal skeleton the bass or perch,
like the greater number of fishes, is more or less enclosed
and protected by an external one, consisting of a beautifully
arranged series of overlapping scales, which afford protection to the underlying organs, and at the same time admit
of great freedom of movement.   These usually consist of a
horny substance, to which lime is sometimes added, and
are peculiar modifications of the skin, something like the
feathers, nails, and hoofs of higher forms.

153. **The air-bladder.**—Naturally a fish's body is heavier
than the water in which it lives, and there are reasons for
thinking that the air-bladder (Fig. 106, *a.bl.*) acts in the

bass and perch and many other fishes as a float to enable
them, without much effort, to remain suspended at a definite level.   By compressing this sac, partly by its own mus.
cles and partly by those of the body-wall, the bulk of the
fish is made less, and it sinks; upon the relaxation of these
same muscles the body expands and rises again.   Deep-set
fishes, when brought to the surface, where the pressure is
relatively slight, are found with their air-bladders so distended that they can not sink again, and the float of surface
fishes would be as useless if they were to be carried into the
depths below, so that such fishes are compelled to keep
within tolerably definite limits of depth.   Morphologically
considered, the air-bladder is a modified or degenerate lung,
and in many fishes it is lost altogether.

154. **Respiration.**—Looking down the throat of the perch,
or any other fish, a series of slits (the gill-openings), usually
four or five in number, may be seen on each side communicating with the exterior.   In the sharks these outer openings are readily seen, but in the bony fishes they open into
a chamber on each side of the head, covered by a bony plate
or gill-cover that is open behind.   On raising these flaps
the gills may be seen composed of great numbers of bright.
red filaments attached to the bars between each slit.   During life the fish may be seen to open its mouth at regular
intervals, and, after gulping in a quantity of water, to close
it again, contracting the sides of the throat to force it out
of the gill-openings and over the gill-filaments to the exterior.   During this process the blood traversing the excessively thin filaments extracts the oxygen from the water and
carries it to other parts of the body.

With this information, let us return to the study of the
three classes of fishes.

155. **The** lancelet (**Branchiostoma**).—The lancelet, sometimes called amphioxus (Fig. 101), the type of the class *Lepto-
cardii*, is a little creature, half an inch to four inches long, in
the different species, transparent and colorless, living chiefly

in sand in warm seas, the ten species being found in as many different regions. A lancelet may be regarded as a vertebrate reduced to its lowest terms. Instead of a jointed back-bone, it has a cartilaginous notochord, running from the head to the tail. A nervous cord lies above it, enclosed in a membranous sheath. No skull is present, and the nerve-cord does not swell into a brain. There are no eyes and no scales. The mouth is a vertical slit, without jaws. There is no trace of the shoulder-girdle (shoulder-blade and collar-bone) or pelvis (hip-bone) from which

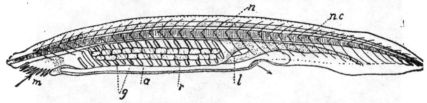

Fig. 101.—The California lancelet (*Branchiostoma californiense*). Twice the natural size. *g*, gills ; *l*, liver ; *m*, mouth ; *n*, nerve-cord ; *nc*, notochord.

spring the paired fins, which, in true fishes, correspond to arms and legs. The circulatory system is fish-like, but there is no heart, the blood being driven about by the contraction of the walls of the vessels. Along the edge of the back and tail is a rudimentary fin, made of fin-rays connected by membrane. In the character and arrangement of its organs the lancelet is certainly like a fish, but in degree of development it differs more from the lowest fish than the fish does from a mammal.

156. **Lampreys (or Cyclostomes).**—The class of lampreys stands next in development (Fig. 102). The notochord gives way anteriorly to a cartilaginous skull, in which is contained the brain, of the ordinary fish type. There are eyes, and the heart is developed, and consists of an auricle and a ventricle. As distinguished from the true fish, the lampreys show no trace whatever of limbs or of the bones which would support them. The lower jaw is wholly wanting, the mouth being a roundish sucking disk. The fins

are better developed, but of the same structure as in the lancelet. There is no bony matter in the skeleton, and there are no scales. The nasal opening is single on the top of the front of the head.

There are about twenty-five species in this class. Some of them, called lampreys, ascend the streams from the sea

FIG. 102.—Lampreys.

in the spring for the purpose of spawning. The young undergo a metamorphosis, at first being blind and toothless. The adults feed mostly on the blood of fishes, which they suck after scraping a hole in the flesh with their rasplike teeth. The others, called hag-fishes, live in the sea and bore into the bodies of other fishes, whose muscles they devour. All are slender, smooth, and eel-shaped.

From their structure and development we suppose that these eel-like forms existed long ago, probably before the more highly developed sharks and bony fishes made their appearance, but it is difficult to determine whether their simple organization is · of such long standing or is not in part the result of semiparasitic habits, or a life spent largely in burrowing. Like the lancelet and other simple chordates, they are of the greatest interest to the zoologist who gains from them some idea of the lowly vertebrate forms that peopled the earth long ago.

157. **True fishes.**—The third class, Pisces or true fishes, to which the shark as well as the bass and perch belong, has a well-developed skeleton, skull, and brain. The lower jaw is developed, forming a distinct mouth, and there is at least a shoulder-girdle and pelvis; although the fins these should bear are not always developed, the general traits are those we associate with the fish. Of the true fishes, there are again several strongly marked groups, usually called sub-classes, two of them wholly extinct. Of these, three chiefly interest us.

158. **The sharks and skates.**—Very early in the life of the sharks (Fig. 103) and skates (*Selachii* or *Elasmobranchii*)

Fig. 103.—Soup-fin shark (*Galeus zyopterus*) from Monterey, Cal.

a notochord appears, similar to that in the lancelet and the lampreys. As growth proceeds its sheath becomes broken up into a series of cartilaginous rings, which thus appear like spools strung on a cord. As the fish grows older these " spools " or vertebræ grow solid, cutting the notochord into little disks, and great flexibility is thus secured. Cartilaginous appendages also grow up and cover the spinal nerve-cord lying above, and give strength to the unpaired fins; the paired fins also have their supports. The shoulder-girdle is placed behind the skull, leaving room for a distinct neck ; strong bars of cartilage bear the gills ; others form jaws to carry the teeth ; and a complex skull protects the brain and sense-organs, which are of a relatively high state of development. Throughout life the skeleton is of cartilage, with perhaps here and there a little bone where greater strength is required. Besides these, there are numerous minor characters which the student will readily find for himself.

The sharks and skates or rays live chiefly in the sea, and some reach an enormous size, the largest of all fishes. Some are very ferocious and voracious; others are very mild and weak, and the development of teeth is in direct proportion to their voracity of habit. In earlier geologic times there were many more species of them than now exist.

159. **The lung-fishes.**—The lung-fishes (Dipnoi) are peculiar forms living in some of the rivers of Australia and the tropical regions of Africa and South America. In these the air-bladder is developed as a perfect lung. During the wet season they breathe like other fishes by means of gills, but as the rivers dry up they burrow into the wet mud and breathe by means of lungs which are spongy sacs of which the air-bladder of other fishes is a degenerate representative. As we shall see, they resemble in this respect the tadpoles and some adult Amphibia (frogs and salamanders). The paired fins are also peculiar in structure, having an elongate jointed axis, with a fringe of rays along its length, a structure almost as much like that of the limbs of a frog as that of a fish's fin. In fact the Dipnoi must be regarded as an ancestral type, an ally of the generalized form from which Amphibia and bony fishes have descended. Only four living species of dipnoans are known, but great numbers of fossil species are found in the rocks.

160. **The bony fishes (Teleostei).**—The bony fishes, or Teleosts, are distinguished by the bony skeleton, the symmetrical tail, and by the development of the air-bladder as a more or less completely closed sac, useless in respiration.

Often this organ is altogether wanting, as in the common mackerel. About twelve thousand kinds of bony fishes are known. The species swarm in every sea, lake, or river throughout the earth, and some form or another among them is familiar to every boy in the land. These fishes are divided into about two hundred families, and these may be arranged in fifteen to twenty orders. As these are mostly distinguished by features of the skeleton, we need not name them here. In Jordan and Evermann's Fishes of North and Middle America, as well as in various other books, the student of fishes can find the characters by which orders may be distinguished.

161. **Sturgeons and garpikes (Ganoidei).**—While the great majority of the typical fishes possess a bony skeleton, there are a few quaint types—the ganoid fishes showing ancient traits. In some of these, as the sturgeon, the skeleton is cartilaginous. In the garpike and bowfin it is long, as in the teleosts. Most of this group are now extinct. At present in this country the ganoids are represented by several species, the best known being the sturgeons. These inhabit the Great Lakes, the Mississippi, and its tributaries; while other species ascend the rivers to spawn. These are the largest fishes found in fresh water, attaining a length of ten or twelve feet, and a weight of five hundred pounds. Their food consists of small plants and animals, which they suck in through their tube-like mouth. The garpikes live in the larger lakes and rivers throughout the East and Mississippi Valley. Their bodies, from three to ten feet in length, according to the species, are covered with comparatively large regularly arranged square scales, and the upper jaw is elongated to form a kind of beak, abundantly supplied with teeth. They are carnivorous, voracious fishes, working great havoc among the more defenseless food-fishes. Equally destructive is the voracious bowfin (*Amia*), a fish useless as food, but of very great interest from its relation to extinct forms.

162. **The catfishes.**—Among the lowest bony fishes we may place the great group to which almost all fresh-water fishes belong. In this group the four vertebræ situated next the head are firmly united, and by means of certain small lever-like bones a connection is formed between the air-bladder and the ear of the fish, which is sunk deep in the skull. The air-bladder thus becomes a sounding organ in the function of hearing. The family of catfishes possesses this structure, and the student should look for it in the first one he catches. The catfishes are remarkable for the long feelers about the mouth, with which they pick their way on the bottom of a pond. There are many kinds the world over. The small ones are known as horned pout or bullhead. In these the dorsal and pectoral fins are armed each with a strong, sharp spine, which is set stiff when the fish is disturbed, and makes them very troublesome to handle. The catfishes have no scales.

163. **The carp-like fishes.**—The still greater carp family includes all the carp, dace, minnows, and chubs. They have the air-bladder joined to the ear, just like the catfish, but they lack the long feelers and the fin spines, while the soft body is covered with scales, and there are no teeth in the mouth. In the throat are a few very large teeth, which the ingenious boy should find. In the sucker family these throat teeth are like the teeth of a comb, and the mouth is fitted for sucking small objects on the river bottom.

164. **The eels.**—In the great order of eels the body is long and slim, scaleless, or nearly so, with no ventral fins. The shoulder-girdle has slipped back from the head, so as to leave a distinct neck, while ordinary fishes have none. Of eels there are very many kinds—some large and fierce, some small as an earthworm; and one kind comes into fresh water.

165. **Herring and salmon.**—In the great order which includes the herring and salmon the vertebræ are all alike, the ventral fins far from the head, and the scales smooth to

the touch. The herring and shad are examples, as also the salmon and trout. Some live in the great depths of the sea, even five miles below the surface. These are very soft in body, being under tremendous pressure. They are inky black—for the sea at that depth seems black as ink—and most of them have luminous spots which give them light in the darkness. Some species have the forehead luminous, like the headlight of an engine. Most of these deep-sea fishes are very voracious, for there is nothing for them to feed on save their neighbors.

166. **The pike, sticklebacks, etc.**—Several small orders stand between these soft-rayed, smooth-scaled fishes and

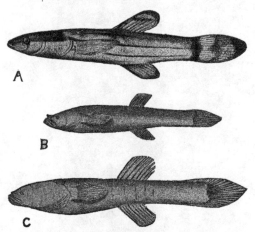

Fig. 104.—The blindfish and its parentage. A, Dismal Swamp fish (*Chologaster avitus*), the ancestor of (B) Agassiz's cave fish (*Chologaster agassizi*) and (C) cave blindfish (*Typhlichthys subterraneus*).

the form, like the perch and bass, which has many spines in the dorsal fin. Among these transitional forms is the pike (Fig. 105)—long, slender, circumspect, and voracious, lying in wait under a lily-pad; the blindfish, which lost its eyes through long living in the streams of the great caves; the stickleback, small, wiry, malicious, and destructive, stealing the eggs and nibbling the fins of any larger fish; the sea-horse, often clinging with its tail to floating

Fig. 105.—The pike (*Esox*). From photograph by R. W. Shufeldt.

seaweed, the male carrying the eggs about in his pocket until they hatch; the mullet, stupid, blundering, feeding on minute plants, crushing them in a gizzard like that of a hen, but withal having soft flesh, good for the table; the flying-fishes, which sail through the air with great swiftness to escape their enemies.

167. **The spiny-rayed fishes.**—In the group of spiny-rayed fishes the ventral fins are brought forward and joined to the shoulder-girdle. The scales are generally rough to the touch, and the head is usually roughened also. There are many in every sea, ranging in size from the Everglade perch of Florida, an inch long, to the swordfish, which is thirty. These are the most specialized, the most fish-like of all the fishes. Leading families are the perch, in the fresh waters, the common yellow perch, familiar to all boys in the Northeastern States; the darters, which are dwarf perches, beautifully colored and gracefully formed, living on the bottoms of swift rivers; the sunfishes, with broad bodies and shining scales, thriving and nest-building in the quiet eddies; the sea-bass of many kinds, all valued for the table; the mackerel tribe, mostly swimming in great schools from shore to shore. After these come the multitude of snappers, grunts, weakfishes, bluefishes, rose-fishes, valued as food. Then follow the gurnards, with bony heads; the sculpins, with heads armed with thorns, the small ones in the rivers most destructive to the eggs of trout; and at the end of the long series a few families in which the spines once developed are lost again, and the fins have only soft and jointed rays. It is a curious law of development that when a structure is once highly specialized it may lose its usefulness, at which point degeneration at once sets in. Among fishes of this type are the codfishes, with spindle-shaped bodies, and the flounders, with flat bodies. The flounders lie on the sand with one side down, and the head is so twisted that the eyes come out together on the side that lies uppermost. This side is col-

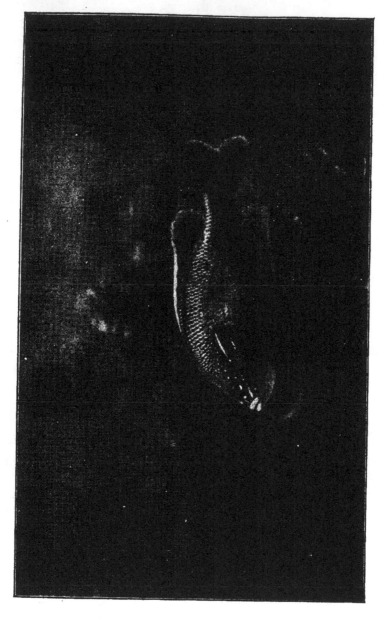

Fig. 106.—Long-eared sunfish (*Lepomis megalotis*).—From photograph by R. W. Shufeldt.

ored like the bottom—sand colored or brown or black—and the under side is white. When the flounder is first hatched, the eyes are on each side of the head, and the animal swims upright in the water like other fishes. But it soon rests on the bottom; it turns to one side, and as the body is turned over the lower eye begins to move over to the other side. Finally, we may close the series with the anglers, in which the first dorsal spine is transformed into a sort of fishing-pole with a bait at the end, which may sometimes serve to lure the little fishes, which are soon swallowed when once in reach of the capacious mouth.

168. **Internal anatomy.**—A few fishes are vegetarians, but the greater number are carnivorous. Some swallow large quantities of sand of the sea-bottom and absorb from it the small organisms living there. Others are provided with beaks for nipping off corals and tube-dwelling worms. Huge plate-like teeth enable others to crush mollusks, sea-urchins, and crabs, and many are adapted for preying upon other fishes. The latter are often able to escape, owing to the presence of numerous spines, sometimes supplied with poison-glands; or their colors are protective, and a vast number of devices are present which enable them with some degree of surety to escape their enemies and capture food.

Usually, without mastication, the food passes into the digestive tract (Fig. 107), which in the main resembles that of the squirrel, but varies considerably according to the nature of the food it is required to absorb. As in other animals, it is usually longer in the vegetable feeders. In most fishes the walls of the canal are pushed out at the junction of the stomach and intestine, to form numerous processes like so many glove-fingers (the pyloric cœca, Fig. 107, *py.c.*), which probably serve to increase the absorptive surface. The same result is obtained in other ways, chiefly by numerous folds of the lining of the canal.

The blood-system is much more complex in the fishes

than in any of the invertebrates. It also differs in its general plan from that of most adult vertebrates, owing to the peculiar method of respiration. In almost every case the vessels returning from all parts of the body unite into one vein leading into the heart, which consists of only one auricle and ventricle (Fig. 107). From the heart the blood

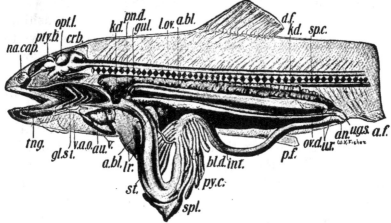

Fig. 107.—Dissection of a bony fish, the trout (*Salmo*). *a.bl.*, air-bladder ; *an.*, anal opening ; *au.*, auricle ; *gl.st.*, gills ; *gul.*, esophagus ; *int.*, intestine ; *kd.*, kidney ; *lr.*, liver ; *l.ov.*, ovary ; *opt.l.*, brain ; *py.c.*, pyloric cœca ; *sp.c.*, spinal cord ; *spl.*, spleen ; *st.*, stomach ; *v*, ventricle.

is forced through the gills, with all their delicate filaments, and now, laden with oxygen and nutritious substances, already absorbed from the coats of the digestive tract, it travels on to all parts of the body, continually unloading its cargo in needy districts and waste matters in the kidneys before returning once more to the heart.

169. **The senses of fishes.**—The habits of fishes indicate that they know considerable of what is going on in the outside world, and their well-developed sense-organs show the degree of their sensitiveness. A share of this information comes through the sense of touch, which is distributed all over the surface of the body, chiefly in the more exposed regions sometimes especially provided with fleshy feelers, like those on the chin of the catfish.

The sense of smell appears to be fairly developed, as is that of hearing; but there is no evidence of a sense of taste. A few fishes chew their food, and may possibly taste it, but there are others that swallow it whole, and in all there are relatively a few nerves going to the tongue or floor of the mouth.

The eyes of most fishes are highly developed, and are of the greatest use at all times. Exceptions to the rule are found in certain species which live in caves or in the dark abysses of the ocean. In some of these the eyes have disappeared amost completely, and the sense of touch becomes correspondingly more acute; in other deep-sea forms they have grown to a large size, enabling them to distinguish objects in the gloom, like the owls and other nocturnal animals. Embedded in the skin of some of these deep-sea fishes, and certain nocturnal ones, are peculiar spots, composed of a glandular substance, which produces a bright glow like that of the fireflies. These may be located on the head or arranged in patterns over various parts of the body, and may serve to light the fish on its way and enable it to see its food to better advantage, or it may act as a lure to many fishes that become victims to their own curiosity. In those fishes which are active most of the time the eyes are located on the sides of the head, and in those which remain at or near the bottom they are turned toward the top; in every case where they can be used to the best advantage.

170. **Breeding habits.**—Among fishes the egg-laying time usually comes with the spring, when the males of several species become more resplendent, and sometimes engage in struggles for their respective mates. In others this ceremony is performed without show of hostility. Some make nests, while others lay their eggs loosely in the water.

In the salmon family, the eggs are laid in cooling waters, as in rivers or brooks. The young make their way downward, often entering the sea. When the young in the sea become mature they emigrate in great companies, and make their way hundreds, perhaps thousands, of miles to the rivers in which they spent their youth. Up these streams they rush in crowds, leaping over waterfalls and rapids, and, dashed and battered on the rocks, many, and in some species all, die from injuries or exhaustion after the breeding season is passed. The eggs, like those of the chubs, suckers, sunfishes, and catfishes, are usually buried in shallow holes in the sand, and the males of most fishes keep a faithful watch over the young until they are able to live in safety. In some of the sticklebacks and in several marine species elaborate nests are composed of grass or seaweeds; some of the catfishes carry the eggs until they hatch in their mouths or else in folds of spongy skin on the under side of the body; in the pipefishes and sea-horses a slender sac along the lower surface of the male acts as a brood-pouch, in which the female places the eggs to remain until developed; and some fishes, such as the surf-fishes and a number of the sharks, bring forth their young alive. On the other hand, the young of many of the herrings, salmon, cod, perch, and numerous other fishes are abandoned at their birth, and fall a prey to many animals, even the parents often devouring their own eggs.

In the former cases, where the young are protected, only a relatively few eggs are produced: where they are abandoned the female often lays many millions. In every case the number of eggs is in direct relation to the chances the young have of reaching maturity, a few out of each brood surviving to perpetuate the race.

171. **Development and past history.**—The eggs of the higher bony fishes are usually small (one-tenth to one-third of an inch in diameter), and the young when they hatch are accordingly little; in the sharks the eggs are larger, the size of a hen's egg or even larger, and the young when born are relatively large and powerful. These differences, however, do not greatly affect the early development, for

in every case the head and then the trunk soon become formed, gills arise, the nervous system appears, which is invariably supported by a skeleton in the form of a gristly rod—the notochord. In the lower forms of fishes this persists throughout life; but in the sharks and skates it becomes replaced in the adult by another and higher type of skeleton, which is much more specialized with the bony fishes.

Those who study the fossils on the rocks tell us that the first fishes were very simple, and many believe that their skeleton, like that of the little growing fish, consisted only of a notochord. Many of these old forms died out long ago, while others gradually changed in one way and another to adapt themselves to their surroundings, the constant need of adaptation having resulted in the multitude of present-day types. Some of the sharks have probably changed relatively only to a slight extent; others, like the garpike, are much more altered; and the bony fishes are far from their original estate, though their development has been rather toward a greater specialization for aquatic life than an advance upward. The little fish in its growth from the egg thus repeats the history of its ancestral development; but as though in haste to reach the adult condition, it omits many important details. Moreover, the record in the rocks is not complete, and we have many things yet to learn of the ancient fishes and their development from age to age to the present day.

CHAPTER XV

In many respects the amphibians—toads, frogs, and salamanders—resemble the fishes, especially the lung-fishes (Dipnoi). The modern amphibians are essentially fishes in their early life, but in developing legs and otherwise changing their bodily form they become adapted for a life on land under conditions differing from those of the fishes. Judging from this class of facts, we may assume that fish-like ancestors, by the development of the lungs, became fitted for a life on land, and that from these the amphibians of our times have been derived.

172. **Development.**—The eggs of the Amphibia are laid during the spring months in fresh-water streams and ponds. They are globular, about as large as shot, and are embedded in a gelatinous envelope (Fig. 108). They are either deposited singly or in clumps, or festooned in long strings over the water-weeds. During the next few days development proceeds rapidly under favorable conditions, resulting in an elongated body with simple head and tail. In this condition they are hatched as tadpoles. As yet they are blind and mouthless, but lips and horny jaws soon appear, along with highly developed eyes, ears, and nose. External fluffy gills arise on the sides of the head, and slits form in the walls of the throat, between which gills are attached, and over which folds of skin develop, as in the fishes. A fin-fold like that of the lancelet or lamprey appears on the tail. The brain and spinal cord, extending along the line of the back, are supported by a gristly notochord, and complete and com-

182

plex internal organs adapt the animal to a free-swimming existence for days to come.

The tadpole is now, to all intents and purposes, a fish— a fact most clearly recognized in its form, method of loco-

Fig. 108.—Metamorphosis of the toad.—Partly after GAGE.

motion, the arrangement of the gills, and the general plan of the circulatory system.

173. **Further growth.**—In the course of the next few weeks hind limbs develop beneath the skin, through which they finally protrude. In the same manner, fore limbs arise at a later date. In position these organs are like the paired fins of fishes, but they are intended for crawling or leaping on land, and are modified in accordance with this need. As in the higher vertebrates, the limbs develop as arms and legs, with long fingers and toes, between which are stretched webs of skin, which serve in swimming.

In the meantime large internal changes are also taking place. The wall of the esophagus has gradually pouched out to form the lungs. They are richly supplied with blood vessels, closely resembling in their general features the lungs of the lung-fishes. The animal now rises to the surface occasionally to gulp in air, and it also continues to breathe by means of gills. At this stage of its existence, therefore, the larva is amphibious (two-living), and we have the interesting example of an animal extracting oxygen from both the water and the air. The diet of the tadpole at this time changes from vegetable to animal substances, and horny teeth give way to the small teeth of the frog, and the digestive system undergoes an entire remodeling to adapt it to its new duties. The young amphibian— whether frog, toad, or salamander—is now a four-legged creature, with well-developed head and tail, with lungs and gills, though the latter are usually fast disappearing, and is rapidly assuming those characters which will fit it for a terrestrial or semiaquatic existence.

174. **The salamanders.**—The changes which now ensue in such a larva in reaching the adult condition are relatively slight in the lower salamanders. The external gills often persist (Fig. 111), the lungs are also functional, and the changes are largely those of increase of size. In the larger number of species the gills disappear more or less completely (Fig. 109), such species often abandoning the water for homes in damp soil or under stones and logs, returning to it only when the time comes for their eggs to be laid. The limbs are always relatively weak, never supporting the body from the ground, but serving in a clumsy way to push it from place to place. In the aquatic forms the tail continues to serve as a swimming organ. In some species the hind legs become rudimentary, or even entirely lacking. A still further modification occurs in a few burrowing species, which move by wrigglings of the body, and are without either pairs of legs.

In geological times many of the salamanders were of great size, several feet in length, and some were enclosed in an armor consisting of bony plates. All now living have the skin naked, and with the exception of the giant species of Japan, three feet in length, and a few similar forms in America, the modern representatives are comparatively

Fig. 109.—Blunt-nosed salamander (*Amblystoma opacum*). Photograph by W. H. Fisher.

feeble and measure their length by inches. Only a few, on account of their bright colors, are particularly attractive, while the others are usually shunned and considered repulsive, chiefly because of their supposed poisonous character, though in reality few animals are more harmless.

175. **Tailless forms.**—In the frogs and toads the metamorphosis which the young undergo is almost as profound as that which takes place with the insects. The gills, together with their blood-vessels, disappear completely. The tail, with its muscles, nerve-supply, and skeleton, is absorbed. The cartilaginous notochord gives way to a jointed back-bone. A skull is developed; numerous bones form in the limbs, affording an attachment for the powerful muscles which make the toad, and especially the frog, expert swim-

mers and leapers, and thus equipped they hereafter lead a wholly terrestrial or semiaquatic life.

176. **Distribution and common forms.**—All the Amphibia are dependent upon moisture. Almost all are hatched and developed in fresh water, and those which leave the water return to it during the breeding season. So we find representatives of the group all over the world having much the same range as the fresh-water fishes. The great majority of the salamanders are confined to the northern hemisphere, but the toads and frogs are almost universally distributed.

Among the salamanders in this country only a relatively few species completely retain their external gills. This is the case with sirens and mud-puppies or water-dogs (Fig. 111), which may occasionally be seen in the clear waters of our lakes and rivers crawling slowly about in search of food, and every now and then rising to the surface to gulp in air. The remainder lose their gills more or less completely, and usually leave the water for damp haunts on land. One of the blunt-nosed salamanders, known as the tiger salamander (*Amblystoma tigrinum*), is found in moist localities in most parts of the United States. Besides these are numerous small species, among them the newts (*Diemyctylus*), ranging widely over the United States, living under logs and stones and feeding upon the small insects and worms inhabiting such situations. In several species of salamanders the lungs disappear with age, and respiration is performed solely through the surface of the skin.

The tailless amphibians are much more abundant and familiar objects than the salamanders, and from the opening of spring until late in the fall they are met with on every hand. With few exceptions the frogs live in or about ponds and marshes, in which they obtain protection in troublous times and from which they derive the store of worms and insects that serve as food. On the other hand, the tree-frogs, as their name indicates, usually abandon the water and repair to moist situations in trees and other vegetation. Their shrill, cricket-like calls are often heard in the summer. The fingers and toes are more or less dilated into disks at their tips, enabling them to climb with considerable facility; and they are further adapted to their surroundings on account of their protective colors. The toads undergo their metamorphosis while very small, and approach the water only at the breeding season. During the day they remain concealed in holes and crevices, but at the approach of evening come out in search of food.

177. **Means of defense.**—The food of the members of this group consists chiefly of small fishes, insect larvæ, snails, and little crustaceans, which are swallowed whole. On the other hand, many Amphibia prey on each other, while most of them are eagerly sought by birds and fishes. Some, as the toads, stalk their food only during the night-time or depend upon their agility to escape their enemies. Others are colored protectively, the markings of the skin resembling the foliage of the earth upon which they rest, and in some species, as the tree-toads, this color-pattern changes as the animal shifts its position. A few species are most brilliantly colored with red, green, yellow, or combinations of these, in striking contrast to their surroundings. They have apparently few enemies, possibly because of an unpleasant odor or taste, and it has been suggested that their gorgeous tints are danger-signals, warning their would-be captors from attempting a second time to devour them. At the same time it is well known that the somber-hued toads emit a milky secretion from the warty protuberance of their skin which is intensely bitter, irritating to delicate skin, and poisonous to several animals.

178. **Skeleton.**—As in all vertebrates, the skeleton of the amphibian first arises as a cartilaginous rod, the notochord, which is afterward replaced by a jointed back-bone, to which the limbs are attached. The back-bone is anteriorly modified into a flat, usually complex, skull. In the salamanders the number of vertebræ is sometimes very large,

and the body correspondingly long and snake-like; but in other cases parts of the vertebræ are reduced in number, and the body is rather short and thick. In the frogs and toads this reduction reaches its culmination, for only nine distinct vertebræ are present, the tail vertebræ, corresponding to those of the salamanders, being represented by a rod-like bone, the urostyle, made of segments grown together.

179. Digestive and other systems.—In its main characters the digestive tract of the amphibian (Fig. 110) resembles

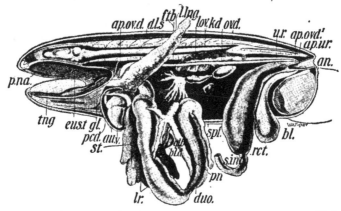

Fig. 110.—Dissection of toad (*Bufo*). *an.*, anal opening; *au.*, auricle; *bl.*, bladder; *duo.*, duodenum; *lng.*, lung; *lr.*, liver; *pn.*, pancreas; *rct.*, rectum; *spl.*, spleen; *st.*, stomach; *v.*, ventricle.

that of the fishes and the squirrel. The mouth is usually large, and the teeth are very small, as in the frog or salamander, or are lacking completely, as in the common toad. In many salamanders the tongue, like that of a fish, is fixed and incapable of movement. In most of the frogs and toads it is attached to the front of the mouth, leaving its hinder portion free, and capable of being thrown over and outward for a considerable distance. In the throat region gill-clefts may persist, but they usually close as the lungs reach their development. The succeeding portions of the canal are comparatively straight in the elongated forms, or

more or less coiled in the shorter species. In some cases
no well-marked stomach exists, but ordinarily the different
portions, as they are shown in Fig. 110, are well defined.

As noted above, the circulation in the tadpole is the
same as in fishes, then lungs arise, and for a time respiration is effected both by gills and lungs, and the circulation resembles in its essential points that of the
lung-fishes. This may continue throughout life, but more
frequently the gills and their vessels disappear, and the
circulation approaches that of the reptiles. In such forms
the heart consists of two auricles and one ventricle. Into
the left auricle pours the pure blood from the lungs; into
the right the impure blood from the body. To some
extent these mix as they are forced into the general circulation by the single ventricle. The amount of oxygen
carried is therefore smaller than in the higher air-breathers,
the amount of energy is proportionately less, and hence it
is that all are cold-blooded and of comparatively sluggish
habits.

In some species of salamanders the lungs may also disappear, and breathing is carried on by the skin, as it is to
a certain extent in all amphibians. In the frogs and toads
lungs are invariably present, and vocal organs are situated
at the opening of the windpipe in the throat. These produce the characteristic croaking and shrilling, which in
many species are intensified through the agency of one or
two large sacs communicating with the mouth-cavity.

Although the brain is small in the amphibians, it is
more complex in several respects than it is in fishes.
The eyes are also usually well developed, but in some of
the cave and burrowing salamanders they are concealed
beneath the skin, and are rudimentary. The ear varies
considerably in complexity in the different species, but in
the possession of semicircular canals and labyrinth resembles that of the fishes. In the frogs and toads, as one may
readily discover, the drum or tympanum is external, ap-

pearing as a smooth circular area behind the eye. Organs
of touch, smell, and taste are likewise developed in varying
degree of perfection.

180. **Breeding-habits.**—While the great majority of amphibians mate in the spring and deposit their eggs in the
water, often to the accompaniments of croakings and pip-
ings almost deafening in intensity, several species, for
various reasons, have adopted different methods. Some of
the salamanders bring forth young alive, and several species
of toads and frogs are known in which the young are cared
for by the parent until their metamorphosis is complete.
In one of the European toads (*Alytes*) the male winds
the strings of eggs about his body until the tadpoles are

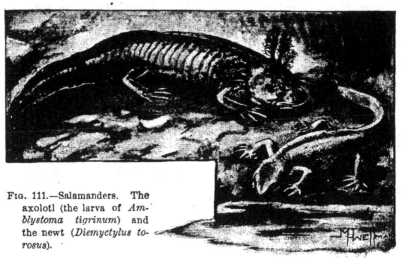

FIG. 111.—Salamanders. The
axolotl (the larva of *Amblystoma tigrinum*) and
the newt (*Diemyctylus to-
rosus*).

ready to hatch ; and in a few species of tree-toads the eggs
are stored in a great pouch on the back of the parent until
the early stages of growth are over. In the Surinam toad
of South America the eggs are placed by the male on the
back of the female, and each sinks into a cavity in the
spongy skin. Here they pass through the tadpole stage
without the usual attendant dangers, and emerge with the
form of the adult.

Sunlight and warmth are apparent necessities for speedy development. Tadpoles kept in captivity where the conditions are generally unfavorable may require years to assume the adult form. As mentioned above, the tiger salamander (*Amblystoma tigrinum*) occurs in most parts of the United States and Mexico. In the East this species drops its gills in early life as other salamanders do, and assumes the adult form, but in the cold water of high mountain lakes, in Colorado and neighboring States, it may never become adult, always remaining as in Fig. 111. This peculiar form is locally known as axolotl. In this condition it breeds. It is thus one of the very few examples of animals whose undeveloped larvæ are able to produce their kind. Owing to this trait it was at first considered a distinct species, and many years elapsed before its relationship to the true adult form was discovered.

CHAPTER XVI

181. **General characteristics.**—In all the reptiles the general shape of the body, and to some extent the internal plan, is not materially different from that seen among the amphibians. In spite of external resemblance the actual relationship is not very close. It appears to be true that ages ago the ancestors of the modern reptiles were aquatic animals, possibly somewhat similar to some of the salamanders; but they have become greatly changed, and are now, strictly speaking, land animals. At no time in their development after leaving the egg do we find them living in the water and breathing by gills. Some species, such as the turtles, lead aquatic or semiaquatic lives, but the modifications which fit them for such an existence render them only slightly different from their land-inhabiting relatives. The skin bears overlapping scales or horny plates, united edge to edge, as in the turtles, enabling them to withstand the attacks of enemies and the effects of heat and dryness. Indeed, it is when heat is greatest that reptiles are most active. In no other class of vertebrates, and very few invertebrates, do normal activities of the body appear to be so directly dependent upon external warmth. In the presence of cold they rapidly grow sluggish, and sink into a dormant state.

As in the case of all animals, habits depend upon structure, and accordingly among the reptiles we find many remarkable modifications, enabling them to lead

192

widely different lives. Nevertheless all are constructed upon much the same plan.

182. **The lizards (Sauria).**—As in the amphibians, especially the salamanders, the body (Fig. 112) consists of a relatively small head united by a neck to the trunk,

FIG. 112.—Common lizard or swift (*Sceloporus undulatus*). Photograph by W. H. FISHER.

which, in turn, passes insensibly into a tail, usually of considerable length. Two pairs of limbs are almost always present, and these exhibit the same skeletal structure as in the amphibians; but in their construction, as in the other divisions of the body, we note a grace of proportion and muscular development which enable the lizards to execute their movements with an almost lightning-like rapidity. The mouth is large and slit-like, well armed with teeth, and the eyes and ears are keen. Scales of various forms and sizes, always of definite arrangement, cover the body. The scales are always colored, in some species as brilliantly as the feathers of birds, and usually harmonize with the surroundings of the animal, enabling it to escape the attacks of its many enemies. Altogether the lizards are a very attractive group of animals. As in the salamanders, the vertebral column is usually of considerable length, but it too presents a lighter appearance and a greater flexibility. Slender ribs are present, and a breast-bone and the girdles which support the limbs. Although more ossified than in the amphibians, the skull still continues to be composed here and there of cartilage. The roof also is yet incomplete, but with the firm plates on the surface of the head ample protection is afforded the small brain underneath. As above mentioned, the limbs are slender and insufficient to support the body, which accordingly rests upon the ground, and by its wrigglings and the pushing of the limbs is borne from place to place. It will be recalled that some of the salamanders living in subterranean haunts and burrowing in the soil have no need of limbs, and the latter have accordingly disappeared. This condition is paralleled by certain species of lizards. The blindworms (which are neither blind nor worms, but true lizards, though snake-like in appearance) are devoid of limbs, as are also the "glass-snakes." In some species the hinder pair arise in early life, but they remain small, and ultimately disappear. In almost all lizards the tail is very brittle, breaking at a slight touch. In such case the lost member will grow again after a time.

183. **The snakes (Serpentes).**—The snakes are characterized by a cylindrical, generally greatly elongated body, in which the divisions into head, neck, trunk, and tail are not sharply defined. As we have seen, this is also true of certain lizards, but the naturalist finds no difficulty in detecting the differences between them. Another peculiarity of the snakes is in the great freedom of movement of the bones

not concerned with the protection of the brain. In the reptiles the lower jaw does not unite directly with the skull, as in the higher animals, but to an intermediate bone, the quadrate, which is attached to the skull. In the snakes these unions are made by means of elastic ligaments. The two halves of the lower jaw are also held

Fig. 113.—Blacksnake (*Bascanion constrictor*). Photograph by W. H. Fisher.

together by a similar band, so that the entire palate and lower jaw are loosely hung together. This enables the snake to distend its mouth and throat to an extraordinary degree, and to swallow frogs and toads but slightly smaller than itself. Where the prey is of relatively small size, the halves of the lower jaw alternate with each other in pulling backward, thus drawing the food down the throat. The food is never masticated. The teeth are usually small and recurved, and serve only to hold the food until it may be swallowed. The latter process is facilitated by the copious secretion of the salivary glands, which become very active at this time.

A further character of the snakes is the absence externally of any trace of limbs. However, in some of the pythons and boas hind limbs are present in the form of small groups of bones embedded beneath the skin and terminating in a claw. There thus appears to be no doubt that the ancestors of the modern snakes were four-footed, lizard-like creatures, which have assumed the present form in response to the necessity of adaptation to new conditions.

More than any other order of vertebrates do the snakes deserve the name of creeping things, and yet their method of locomotion enables them to crawl and swim with a rapidity equal to that of many of the more highly developed animals. This depends chiefly upon certain peculiarities of the skeleton, which consists merely of a skull, vertebral column, and ribs. The vertebræ, usually two or three hundred in number, are united together by ball-and-socket joints, and each attaches by similar joints a pair of slender ribs. These in turn are attached to the broad outer plates upon which the body rests, and the whole system is operated by a powerful set of muscles. Upon the contraction of the muscles the ventral plates are made to strike backward upon the ground or other rough surface, which drives the body forward. Also, the ribs may be made to move backward and forward, and the snake thus progresses like a centiped or " thousand-legs."

184. **The turtles (Chelonia).**—In many respects the turtles are the most highly modified of all the reptiles. The body (Fig. 114) is short and wide and enclosed in a shell or heavy armor, consisting of an upper portion, the carapace, and a flat ventral plate, the plastron. The shape of the carapace varies greatly from a low, flat shield to a highly vaulted dome, remaining cartilaginous throughout life, as in the soft-shelled turtles, or becoming bony and of great strength. The two portions of the shell form a box-like armor through whose openings may be extended the head, tail, and limbs. As a means of protection the turtle may retract these organs within the shell. The head is generally

.thick-set and muscular, and provided with horny jaws entirely destitute of teeth, like those of the birds. The limbs also are usually short and thick and variously shaped, and adapted for aquatic or terrestrial locomotion. The number of vertebræ in the body and tail are relatively few, and the thick and heavy body is devoid of the elements of grace and agility of movement characteristic of the other reptiles. On the other hand, the former enjoy a freedom from the attacks of enemies not accorded to animals in general.

At first sight the appearance of a turtle does not indicate a close relationship to the other reptiles, but a more

Fig. 114.—Box-turtle (*Terrapene carolina*).

careful examination, and especially of their development, discloses a remarkable resemblance. The head, tail, and limbs are essentially similar to those of the lizards, but in the trunk region peculiar modifications have taken place. The ribs at first separate, as in other animals, flatten greatly, and unite with a number of bones embedded in the skin, thus forming one great plate overlying the back of the animal. About the circumference of the shield other dermal or skin-bones are added, which increase the area of the carapace, and at the same time still others have arisen and united on the ventral surface to form the plastron. In this process the shoulder- and hip-girdles which attach the limbs come to be withdrawn into the body, and we have the curious example of an animal enclosed within its back-bone and ribs. This is even more the case with the box-turtles (Fig. 114), common in the eastern United States, whose ventral plate is hinged so that after the limbs, head, and tail have been withdrawn it may be made to act like a lid to completely enclose the fleshy parts of the body.

Scales and horny plates are present, as in other reptiles, the former covering all parts of the body except the carapace and plastron, which support the plates. In nearly all species the latter are of considerable size, and in the tortoise-shell turtles are valuable articles of commerce. They also are sculptured in a fashion characteristic of each species, and may, like the colors of other animals, render them more like their surroundings, and consequently inconspicuous.

185. **Crocodiles and alligators (Crocodilia).**—The alligators (Fig. 115) and crocodiles are much more complex in structure than the lizards, though their general form is much the same. The body is covered with an armor of thick bony shields and horny scales. These, along the median line, are keeled, and extending along the length of the laterally compressed tail form an efficient swimming organ and rudder. The mouth is of large size, and is bounteously supplied with large conical teeth, which are set in sockets in the jaw, and not fused with it, as in many of the lizards. The nose and ears may be closed by valves to prevent the entrance of water, and a similar structure blocks its passage beyond the throat while the mouth is open. When large animals, such as hogs or calves, are captured as they come to drink, these devices enable the alligator or crocodile to sink with them to the bottom and hold them until drowned. The limbs, short and powerful, are efficient organs of locomo-

tion on land, and together with the general shape of the body, are also well adapted for swimming.

Fig. 115.—Alligator (*Alligator mississippiensis*).

186. **Distribution of the lizards.**—In a general way the number of reptiles is greatest where the temperature is highest. The tropics therefore abound in species, often of large size, and usually of bright coloration. As one travels northward the numbers rapidly diminish, their size is smaller, and the tints less pronounced. In all probability not less than four thousand known reptiles exist, whose haunts are of the most varied description.

In North America the lizards are almost exclusively confined to the southern portions, only a very few species extending up to the fortieth parallel. Among these the skinks (*Eumeces*) are most widely distributed. The blue-tailed skink is probably the most familiar, a small lizard eight or ten inches in length, dark green with yellowish streaks and a bright-blue tail. On sunny days it may sometimes be seen darting about on the bark of trees in search of insects, upon which it feeds.

One of the most familiar lizards in this country is the "glass-snake," found burrowing in the drier soil of the southern half of the United States east of the Mississippi.

Both pairs of limbs are absent, but by wriggling movements of the body this lizard is able to force its way through light soil with considerable rapidity. It is a matter of some difficulty to secure entire specimens, for with other than the gentlest handling the tail severs its connection with the body, as the vertebræ in this portion are extremely brittle. This peculiarity, together with its shape, has given it the popular name of glass-snake. Many species of lizards will thus detach the tail. a habit which is a means of protection, enabling the animal to scamper away into a place of safety while its enemy is concerning itself with the detached member. Later on a new tail develops, though usually of a less symmetrical form.

187. **Horned toads.**—The horned toads (*Phrynosoma*) are lizards peculiar to the hot, sandy deserts and plains of

FIG. 116.—Gila monster (*Heloderma suspectum*). One-third natural size.

Mexico and the western United States. The body is comparatively broad and flat, almost toad-like, and is covered with scales and spines of brownish and dusky tint, so like dried sticks and cactus spines in form and color as to render them difficult of detection. In captivity they readily adapt themselves to their new surroundings, become tame, and feast on flies, ants, and other insects, which they capture by the aid of their long tongue. The horned toads are perfectly harmless creatures, but when irritated sometimes perform the remarkable feat of spurting a stream of blood from the eye toward the intruding object for a distance of several inches. This has been regarded by some as a zoological fable; but there are many who have watched the horned toad in its natural state and in captivity, and they assure us that it is a fact.

In the hot deserts of Arizona and Sonora is another peculiar species of lizard known as the Gila monster (*Heloderma*) (Fig. 116), having the distinction of being the only poisonous lizard known. Further protection is afforded by bony tubercles on the head and by scales over the remainder of the body, all of which are colored brown or various shades of yellow, giving the animal a peculiar streaked and blotched appearance.

188. **Distribution of the snakes.**—The snakes are much more common than the lizards. All over the United States one meets with them, especially the garter- or water-snakes. Of less wide distribution are the black-, grass-, and milk-snakes, and a number of less known species, all of which are perfectly harmless and often make interesting pets. Some of them when cornered show considerable temper, flatten the head and hiss violently, and imitate poisonous forms, but venomous snakes are comparatively few in number in northern and eastern United States. In the southern portions of the country they become more abundant. Along the streams and in the swamps the copperheads, and especially the water-moccasins, often lie in wait for frogs and fish. Both these species are especially dreaded, as they strike without giving any warning sound, but the name and bad reputation of the moccasin is often, especially in the South, transferred to perfectly harmless water-snakes. On higher ground are the rattlesnakes (*Crotalus*), once

abundant but now in many regions well-nigh exterminated. In these species the tail terminates in a series of horny

FIG. 117.—Diamond-rattlesnake (*Crotalus adamanteus*). Photograph by W. H. FISHER.

rings that produce a buzzing sound like that of the locust when the tail is rapidly vibrated.

189. Distribution of the turtles.—The turtles are perhaps somewhat less dependent upon warmth than other reptiles, yet they too delight to bask in the sunshine, and soon grow sluggish in its absence. In all our fresh-water streams and ponds they are familiar objects, and several species extend up into Canada. Among the turtles the soft shell, the painted and the snapping turtles have the widest distribution, scarcely a good-sized stream or pond from the Gulf of Mexico to Canada, and even farther north, being without one or more representatives. All are carnivorous and voracions, and the snapping turtles are especially ferocious, and "for their size are the strongest of reptiles." In the woods and meadows the wood-tortoise and box-turtles are occasionally met with, and at sea several turtles exist, some of them of great size. Among these is the leather-turtle, found in the warmer waters of the Atlantic, lazily floating at the surface or actively engaged in capturing food. They attain a length of from six to eight feet, and a weight of over a thousand pounds, and are sometimes captured for food when they come ashore to bury their eggs in the sand. By this same method the loggerheads, the hawkbills, and the common green turtles are also captured in considerable numbers. These are of smaller size, and the second named is of considerable value, as the horny plates cover-

FIG. 118.—Hawkbill turtle (*Eretmochelys imbricata*).

ing the shell furnish the tortoise-shell of commerce. These plates are removed after the animal is killed, by soaking in warm water or by the application of heat.

190. **Food and digestive system.**—Some reptiles, among which are a number of species of lizards and the box- and green turtles, are vegetarians, but the great majority are

carnivorous, and usually very voracious. The lizards especially devour large quantities of insects and snails, together
with small fishes and frogs. The latter figure largely in
the turtle's bill of fare, and in that of the snakes, which
also capture birds and mammals. On the other hand, many
of the reptiles prey upon one another; and they are the
favorite food of hawks and owls and numerous water-birds,
of skunks and weasels and many other animals, which look
for them continually. Many of the turtles, owing to their
protective armor, and the snakes because of their poisonous bite or great size and strength, are more or less ex-
empt, but this is not true of their eggs and young. The
smaller species depend upon keenness of sense, agility, and
inconspicuous tints. These latter may undergo changes
according to the character of the surroundings, but usually
only to a slight extent. The chameleons of the tropics
and a similarly colored green lizard on the pine-trees in
the Southern States are able to change with great rapidity
from green, through various shades, to brown.

191. **Respiration and circulation.**—While still in the egg
the young lizard develops rudimentary gills, and thus bears

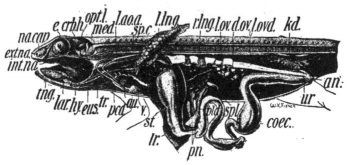

Fig. 119.—Dissection of lizard (*Sceloporus*). *an.*, anal opening; *au.*, auricle; *crb.h.*,
brain; *coec.*, intestine; *kd.*, kidney; *l.lng.*, left lung; *lr.*, liver; *pn.*, pancreas;
sp.c., spinal cord; *spl.*, spleen; *st.*, stomach; *v.*, ventricle of heart.

evidence to the fact that its distant ancestors were aquatic;
but before hatching they disappear, and lungs arise, which

remain functional throughout life. Corresponding to the shape of the body, these are usually much elongated and ordinarily paired (Fig. 119, *l.lng.*). The snakes are peculiar in having the left lung rudimentary or even lacking completely, while the right one becomes greatly elongated and extends far back into the body. In nearly all the reptiles the amount of oxygen brought into the lungs is relatively large and the activity of the animal is proportionately great. The circulation of reptiles shows an advance beyond that of the Amphibia. As in the latter, there are two distinct auricles; but the chief difference arises from the fact that the ventricle is more or less divided by a partition which to a considerable degree prevents the blood returning from the lungs from mixing with the impure blood as it returns from its journey over the body. In the crocodiles and alligators the partition is complete, and the circulation thus approaches close to that of the higher animals.

192. **Hibernation.**—Attention has already been called to the fact that reptiles are very susceptible to cold, rapidly growing less active as the temperature lowers. When winter comes on they seek protected spots, and either alone or grouped together hibernate. The various activities of the body during this period are at very low ebb. The blood barely circulates, breathing is imperceptible, and stiff and insensible to the world about them they remain until the warmth again stirs them to their former activity. Some of our common turtles must also pass a somewhat similar sleep while embedded far down in the mud during the disappearance of the ponds in summer. At such times no food is taken, but owing to their loss in weight it is probable that a slow consumption of the body supplies the small amount of necessary energy.

193. **Nervous system and sense-organs.**—At first sight one is struck with the small size of the brain of fishes, Amphibia, and reptiles. Their intelligence likewise is at low

ebb. Almost all the movements and operations of the body appear to be carried on by the animal with little apparent thought. Their acts, like most of the animals below them, are said to be instinctive; yet they are sufficiently well done to enable the animal to procure its food, avoid its enemies, and lead a successful life. As is true of other animals, the ability of the reptile to cope with its surroundings depends to a great extent upon the keenness of one or all of its organs of special sense. In the reptiles the sense of sight is perhaps sharpest, but there is considerable variation in this respect. Movable eyelids are present in most lizards, together with a third, known as the nictitating membrane, a thin, transparent fold located at the inner angle of the eye, over which it is drawn with great rapidity. Snakes have no movable eyelids, hence the eye has a peculiar stare. Furthermore, their sense of sight, except in a few tree-dwelling species, appears to be defective, the majority depending largely upon the sense of touch.

In all the vertebrates a very peculiar organ known as the pineal gland or eye is situated on the roof of the brain. In several lizards its position is indicated by a transparent area in one of the plates of the head, and by an opening in the bones of the roof of the skull. In young reptiles, and especially in one of the New Zealand lizards (*Hatteria*, Fig. 120), its resemblance to an eye is decidedly striking. Lens, retina, pigment, cornea, are all present much as they are in some of the snails, but they finally degenerate more or less as the animal reaches maturity. It is a general belief that it represents the remnant of an organ of sight, a third eye, which looked out through the roof of the skull in some of the ancient vertebrates.

With the possible exception of the few species of reptiles which produce sounds, probably to attract their mate, the sense of hearing is not particularly well developed. The senses of smell and taste are also comparatively feeble. The latter sense is located in the tongue, which is also popularly supposed to serve for the purpose of defense, and that it is in some way related to the poison-glands. This, however, is an error. The tongue is used primarily as an organ of

Fig. 120.—Tuatera (*Sphenodon punctatus*).

touch, and in snakes especially it is almost continually darted in and out to determine the character of the animal's surroundings.

194. **Egg-laying.**—The eggs of the reptiles are relatively large and enclosed in a shell like a bird's egg, the shell, however, being leathery rather than made of lime. These are deposited in some warm situation, and generally left to themselves to hatch. Under stones, logs, and leaves, or buried lightly in the soil, are the positions most frequently chosen by the lizards and snakes. The turtles almost invariably select the warm sand at the edge of the water, and after scooping a hole lay numerous spherical eggs, usually at night. The alligators lay upward of a hundred eggs about the size of those of a goose, and guard them jealously until and even after they hatch. On the other hand, the young of many lizards and snakes are born alive, the eggs being hatched within the body.

Many reptiles are surprisingly slow in attaining maturity, and live to an age attained by few other animals. It is a well-known fact that turtles live fully a hundred years, and probably the same is true of the crocodiles and alligators and some of the larger snakes. Their enemies are few, and death usually results when the natural course is run.

Throughout life all reptiles periodically shed their skin, as birds do their feathers and mammals their fur. In the snakes and some of the lizards the skin at the lips loosens, and the animal gradually slips out of its old slough, bright and glossy in the new one which previously developed. In the others the old skin hangs on in tatters, gradually coming away as they scamper through the grass.

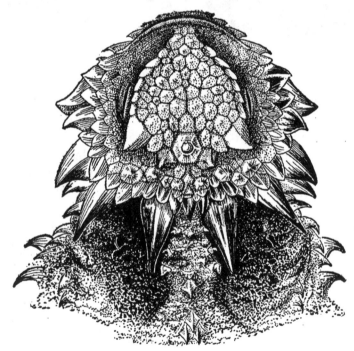

Fig. 121.—Head of the lizard, or "horned toad" (*Phrynosoma blanivillei*), showing the translucent pearly scale covering the pineal eye.—From nature, by W. S. ATKINSON.

CHAPTER XVII

THE BIRDS

195. Characteristics.—Birds form one of the most sharply defined classes in the animal kingdom, and the variations among the different species are relatively small. "The ostrich or emu and the raven, for example, which may be said to stand at opposite ends of the series, present no such anatomical differences as may be found between a common lizard and a chameleon, or between a turtle and a tortoise," and these we know to be relatively slight.

In many respects the birds resemble the reptiles, and long ago in the world's history the relationship was much closer than now, as we know from certain fossil remains in this country and in Europe. One of the earliest of these fossil birds, the Archæopteryx, is a most remarkable combination of bird and lizard. Unlike any modern bird, the jaws were provided with many conical reptile-like teeth. The wings were rather small, and the fingers, tipped with claws, were distinct, not grown together, as in modern birds. The tail was as long as the body, and many-jointed, like a lizard's, each vertebra carrying two long feathers. The bird was about the size of a crow, and it probably could not fly far. Other ancient types have been discovered—principally sea-birds—many of which existed when the Pacific extended over the region now occupied by the Rocky Mountains. These were all of the same generalized type, intermediate between reptile and bird. This fact leads us to the belief that birds descended from reptilian ancestors, and in becoming more perfectly adapted for an aerial life have developed into our modern forms.

In the modern birds the most important peculiarities, those which separate them from all other animals, are correlated with the power of flight. The body is spindle-shaped, for readily cleaving the air. The fore limbs serve as wings. The hind limbs, supporting the weight of the body from the ground, are usually well developed. A series of air-chambers usually exists in powerful fliers. This serves a purpose analogous to that of the air-bladder of a fish, giving buoyancy. But the most characteristic mark of a bird, as above stated, is its feathers, universally present and never found outside the class. Like the scales of lizards, and probably derived from similar structures, they are of different forms, and serve a variety of purposes. The larger ones, with powerful shafts, and forming the tail, act as a rudder. Those of the wings give great expanse with but little increase in weight, and are so constructed that upon the down-stroke they offer great resistance to the air, and push the bird forward, while in the reverse direction the air slips through them readily. In flight these movements of the wing may be too rapid for us to follow, as in the humming-birds, though they are usually much slower, two to five hundred a minute in many powerful fliers, such as the ducks, and frequently long-continued enough to carry them many hundreds of miles at a single flight. The remaining feathers are soft and downy, giving roundness to the body and enabling it to cleave the air with greater ease, and, being poor conductors of heat, they aid in keeping the body at the high temperature characteristic of birds. In most birds the body is not uniformly clothed in feathers. Naked spaces, usually hidden, intervene between the feather tracts, and on the feet and toes scales exist.

196. **Molting.**—As we all know, the growth of feathers, unlike that of hair and nails, is limited, and after they have become faded and worn out they are shed, and new ones

arise to take their place. This process of molting is usually accomplished gradually, without diminishing the powers of flight; but in the ducks and some other birds all the wing- and tail-feathers drop out simultaneously, leaving the bird to escape its enemies by swimming and diving. The molting-process usually takes place in the fall, after the nesting and care for the young is over, and often when the need for a heavy winter coat commences to be felt. Many birds also don what are called courting colors, ruffs, crests, and highly colored patches, in the spring, previous to the mating season, doubtless for the purpose of attracting or impressing their mates. In other cases the change appears to be related to the bird's surroundings. A most beautiful example of this is the ptarmigans—grouse-like birds living far to the north. During winter they are perfectly white and are almost invisible against the snow; but in the spring, as the snow disappears, the white feathers gradually fall out and new ones arise. The latter so harmonize "with the lichen-colored stones among which it delights to sit, that a person may walk through a flock of them without seeing a single bird."

There are also numerous birds, chiefly those that go in flocks, which possess what are known as color-calls or recognition-marks. These may consist of various conspicuous spots or blotches on different parts of the head or trunk, such as we see in the yellowhammer or meadow-lark; or one or more feathers of the wings or tail may be strikingly colored, as in many sparrows and warblers. During the time the bird remains at rest these usually are concealed under neighboring feathers, but during flight they are strikingly displayed. It may possibly be true, as many have urged, that these color-signals are for the purpose of enabling various members of the flock to readily follow their leader; but this and many other interesting questions regarding the color of birds and other animals have not yet received final answers.

In very many animals, fishes as well as birds, the tints on the under side of the body are usually relatively light colored, shading gradually into a darker tint above. This is in all probability a protective device, as was recently shown by Mr. A. H. Thayer, an American artist. His experiments show that the light from above renders the back less dark, and that the shadow beneath is neutralized by the light color. The bird thus appears uniformly lighted, and this effect, together with streaks and blotches, renders them invisible at surprisingly short distances.

197. Skeleton.—Turning now to the internal organization of birds, we find many points in common with other vertebrates, especially the reptiles, but many interesting modifications are also·present that adapt them for flying and for collecting their food. According to the nature of the food, the beak may have a great variety of forms. The skull may be thick and heavy, or thin and fragile, but these are matters of proportion of the various parts possessed by all birds. The neck also is of differing length; but it is in the trunk region that the greatest changes have arisen, as we may see in any of our ordinary birds. For example, the vertebræ of this part of the body are more or less fused together into rigid framework, to which are attached the ribs that in turn unite with the breast-bone. In the fliers the latter bears a vertical plate or keel, to which the great muscles that move the wings are attached. The tail consists, like that of the old-fashioned birds, of several vertebræ, but these are of small size and fused together into a little knob that supports the tail-feathers. The fore limbs are used for flight, but there are the same bones that exist in the fore limbs of other vertebrates—one for the upper arm, two for the lower, a thumb carrying a few feathers, and known as the bastard wing, and indications of several bones that form the hand. In the hind limb the resemblance is equally apparent, though its different parts are of relatively large size to support the body. It is interesting to note that the knee has been drawn far up into the body, and that the joint above the foot is in reality the ankle.

We thus see that the bird's skeleton presents the same general plan as that of the lizard, for example; but in order to combine the elements of strength, lightness, and compactness essential to successful flight, it has been necessary to remodel it to a considerable degree.

198. Other internal structures.—The lungs of birds consist of two dark-red organs buried in the spaces between the ribs along the back. Each communicates with extensive thin-walled air-sacs extending into the space between the

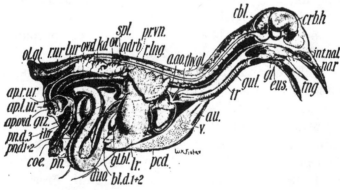

Fig. 122.—Anatomy of a bird. *au.*, auricle; *cbl.* and *crb.h.*, cerebellum and cerebral hemispheres (divisions of the brain); *duo.*, intestine (with portion removed); *giz.*, gizzard; *kd.*, kidney; *r.lng.*, lung; *tr.*, trachea or windpipe; *vent.*, ventricle.

various organs, and in many birds of flight extending into the bones of the body, decreasing their specific gravity. "The enormous importance of this feature to creatures destined to inhabit the air will be readily understood when we learn that a bird with a specific gravity of 1.30 may have this reduced to only 1.05 by pumping itself full of air."

As we know, air is taken into the body in order that the oxygen it contains may combine with the tissues of the body to liberate the energy necessary for the work of its

life. The life of birds is at high pressure, hence their need
of much oxygen. They habitually breathe deeper breaths
than other animals. The air passing into the body trav-
erses the entire extent of the lung on its way back to the
air-sacs, with the result that large quantities of oxygen are
taken into the body. This is distributed by a circulatory
system of a more highly developed type than in any of the
preceding groups of animals. The ventricles of the heart
no longer communicate with each other, and the pure and
impure blood never mingle. Furthermore, the beating of
the heart is comparatively rapid, rushing the oxygen as
fast as it enters the blood to all portions of the body. The
result is that everywhere heat is being generated, so necessary to life and activity.

In the lower animals no special means are employed to
husband the energy thus produced, but in the birds the
body is jacketed in a non-conducting coat of feathers which
prevents its dissipation. For this and other reasons the
birds, summer and winter, maintain an even and relatively
high temperature (102°–110°). Like the mammals, birds
are warm-blooded animals, full of energy, restlessly active
to an extent realized in few of the cold-blooded animals.

199. **Digestive system.**—This life, at high pressure, demands a relatively large amount of food to make good the
losses due to oxidation. The appetites of some growing
birds is only satiated after a daily meal equal to from one
to three times their own weight, and after reaching adult
size the amount of daily food required is probably not less
than one-sixth their weight. The nature of the food is
exceedingly varied, and the digestive tract and certain accessory structures are obviously modified in accordance
with it. The beak, always devoid of teeth in the living
form, varies extremely according to the work it must perform. The same is true of the tongue, and many correlated
modifications exist in the digestive apparatus. In the
birds of prey and the larger seed-eating species, such as the

pigeons and the domestic fowls, the esophagus dilates into a crop, in which the food is stored and softened before being acted upon by the gizzard. The latter is the stomach, provided with muscular walls, especially powerful in the seed-eaters, and with an internal corrugated and horny lining which, in the absence of teeth, serves to crush the food. In some species, such as the domestic fowls and the pigeons, this process is aided by the grinding action of pebbles swallowed along with the food. The remaining portions, with pancreas and liver, vary chiefly in length, and are sufficiently shown in Fig. 122 to require no further description.

200. **Nesting-habits.**—A few birds, such as the ostriches and terns, merely scoop a hollow in the earth, and make no further pretense of constructing a nest. On the other hand, some birds, such as the humming-birds and pewees, build wonderful creations of moss, lichens, and spider-webs, lining it with down, and concealing it so skilfully that they are not often found. Every bird has its own particular ideas as to the fitness of its own nest, and the results are remarkably different, and form an interesting feature in studying the habits of birds. Usually the female takes upon herself the choice of the nest and its construction; but these duties are in some species shared by the male. After the eggs are laid, the male may also aid in their incubation, or may carry food to the female. In other species—for example, the pigeons and many sea-birds—the parents take turns in sitting upon the eggs and in the subsequent care of the young. Finally, there are certain birds, such as the cuckoo and cowbirds, which take advantage of the industry of other species and deposit an egg or two in the nests of the latter. All the work of incubation and care of the young is assumed by the foster-parents, which sometimes neglect their own offspring in their desperate attempts to satisfy the appetites of the rapidly growing and unwelcome guests.

The eggs of birds are relatively large, and are often delicately colored. In some species the blotches and streaks of different shades are probably protective, as in the plovers and sandpipers, whose eggs blend perfectly with their surroundings, but many other cases exist not subject to such an explanation.

The young require a high degree of heat for their development, and this is usually supplied by the parent. In a very general way the length of sitting, or incubation, is proportional to the size of the egg, being from eleven to fourteen days in the smaller species, to seven or eight weeks in the ostriches. Before hatching, a sharp spine develops on the beak, and with this the young bird breaks its way through the shell. Among the quails, pheasants, plovers, and many other species, the young are born with a covering of feathers, wide-open eyes, and the ability to follow their parents or to make their own way in the world. Such nestlings are said to be *precocial*, in distinction to the *altrical* young of the more highly specialized species, such as the sparrows, woodpeckers, doves, birds of prey, and their allies, which are born helpless and depend for a considerable time on the parents for support.

Some of the owls, crows, woodpeckers, sparrows, quails, etc., remain in the same localities where they are bred. They are resident birds. Most kinds of birds, at the approach of winter, migrate toward the southern warmer climes, some species traveling in great flocks, by day or night, and often at immense heights. In some cases this movement appears to be directly related to the food-supply; but there are many apparent exceptions to such a theory, and it is possible that many birds migrate for other reasons. Certain species migrate thousands of miles, along fairly definite routes, the young, sometimes at least, guided by the parents, which in turn appear to remember certain landmarks observed the year before. Sea-birds, in their journeys northward or southward, keep alongshore, occasionally veering in to get their bearings or to rest, especially in the presence of fogs.

201. **Classification.**—Most zoologists make two primary divisions of the living types of birds—those like the ostrich with flat breast-bones, and the other the ordinary birds, in which the breast-bone has a strong keel for the attachment of the powerful muscles used in flight. This distinction is not of high importance, but we may use it as a convenience in the description of a few typical forms belonging to several orders into which these two divisions are subdivided.

202. **The ostriches, etc. (Ratitæ).**—From specimens introduced or from pictures we are doubtless familiar with the ostriches and with some of their relatives. The African ostrich (*Struthio camelus*, Fig. 123) is the largest of living birds, attaining a height of over seven feet, and is further characterized by a naked head and neck, two toes, and fluffy, plume-like feathers over parts of the body. They are natives of the plains and deserts of Africa, where they travel in companies, several hens accompanying the male. When alarmed, they usually escape by running with a swiftness greater than that of the horse, but if cornered they defend themselves with great vigor by means of their powerful legs and beaks. Their food consists of insects, leaves, and grass, to which is added sand and stones for grinding the food, as in the domestic fowl. The American ostriches or rheas, are smaller ostrich-like birds, living on the plains of South America. Their habits are essentially the same as those of the African species.

203. **The loons, grebes, and auks (Pygopodes).**—The birds in this and some of the following orders are aquatic in their habits. All have broad, boat-like bodies, which, with the thick covering of oily feathers, enables them to float without effort. The legs are usually placed far back on the body—a most favorable place for swimming, but it renders such birds extremely awkward on land. The grebes are preeminently water-birds. The pied-billed grebe or dab-

Fig. 123.—African or two-toed ostrich (*Struthio camelus*). Photograph by William Graham.

chick (*Podilymbus podiceps*), for example, found abundantly on the larger lakes and streams throughout the
United States, captures its food, sleeps, and breeds without leaving the water. The loons living in the same situations as the dabchick are also remarkable swimmers and
divers. Of the three species found in this country, the
common loon or diver (*Gavia imber*) attains a length of
three feet, and is otherwise distinguished by its black
plumage, mottled and barred with white, which is also the
color of the under parts. The auks, murres, and puffins
are marine, and, like their inland relatives, are expert
swimmers and divers, strong fliers, and spend much of their
time on the open sea. During the breeding-season they
assemble in vast numbers on rugged cliffs along the shore,
and lay their eggs on the bare rock or in rudely constructed
nests.

204. **The gulls, terns, petrels, and albatrosses** (Longi-
pennes).—The birds belonging to this group are among the
most abundant along the seacoast, and several species make
their way inland, where they often breed. All are characterized by long, pointed wings and pigeon or swallow-like
bodies, which are carried horizontally as the bird waddles
along when ashore. Many are excellent swimmers and
powerful fliers, especially the petrels and albatrosses, which
sometimes travel hundreds of miles at a single flight.

The gulls are abundantly represented along our coasts,
where they frequently associate in companies, usually resting lightly on the surface of the water, or wheeling lazily
through the air on the lookout for food. The terns are
of lighter build than the gulls and are more coastwise in
their habits, and are further distinguished by plunging like
a kingfisher for the fishes on which they live. Both the
gulls and terns breed in colonies, every available spot over
acres of territory being occupied by their nests, which are
usually built of grass and weeds placed on the ground.

The petrels and albatrosses are at home on the high

seas, rarely coming ashore except at the breeding-season. Some species of the former are abundant off our shores, especially the stormy petrel (*Procellaria pelagica*) or Mother Carey's chickens (*Oceanites oceanicus*), which are often seen winging their tireless flight in the wake of ocean vessels. Among the dozen or so albatrosses few reach our shores. The wandering albatross (*Diomedea exulans*), celebrated in story and as the largest sea-bird (fourteen feet between the tips of its outstretched wings), is an inhabitant of the southern hemisphere, and only rarely extends its journeys to more northern regions.

205. **Cormorants and pelicans (Steganopodes).**—The cormorants and pelicans are comparatively large water-birds

Fig. 124.—White pelicans (*P. erythrorhynchus*) and whooping-crane (*Grus americana*). Photograph by W. K. Fisher.

usually abundant along the seashore and in many sections of the United States. The cormorants or shags are glossy black in color, with hooked bills, long necks, and short wings, which give them a duck-like flight. The much larger pelicans (Fig. 124) are at once distinguished by long bills, from which is suspended a capacious membranous sac. All these birds are sociable in their habits, breeding, roosting, and fishing in great flocks. Their food consists of fishes, which the shags pursue under water and capture in their hooked beaks; while the pelicans, diving from a considerable height or swimming rapidly on the surface, use their pouches as dip-nets. The nests, usually built of seaweed or of sticks, are placed on rocky cliffs or on the ground in less elevated places.

206. **Ducks, geese, and swans (Lamellirostres).**—The birds of this order, with their broad, flat, serrated beaks, short legs, and webbed feet, are well known, for in a wild or domesticated state they extend all over the earth. All are excellent swimmers, many dive remarkably well, and are strong on the wing. While a considerable number breed within the United States, their nesting-grounds are generally farther north, and in the early spring it is not unusual to see them migrating in flocks from their warmer winter homes. Among the ducks, the mergansers, mallards (from which our domestic species have been derived), the teals, and the beautiful wood-duck remain with us the year round, dwelling on quiet streams and shallow ponds, living on fish, Crustacea, and seeds. In the more open waters of the larger lakes and along the seacoast we find the canvasback, the scaup-ducks, and the eiders which supply the famous down of commerce. Of the few species of geese which inhabit the United States, the Canada goose (*Branta canadensis*) is perhaps the most familiar. During their migrations to the nesting sites they fly in V-shaped flocks, their "honks" announcing the opening of spring. The brant (*B. bernicla*) is also common in the eastern part of the country, where it, like its relations, lives on vegetable substances entirely. The swans are familiar in their semi-

domesticated state, but the two beautiful wild swans found in this country are rarely seen.

207. The herons and bitterns (Herodiones).—The herons and bitterns are also aquatic in their habits, but, unlike the swimming-birds, they seek their food by wading. Adapting them for such an existence, the legs and neck are usually very long, and the bill, longer than the head, is sharp and slender. Among the relatively few species in the United States, the great blue heron (*Ardea herodias*) is widely distributed, and may often be seen standing motionless in some shallow stream on the lookout for fish, or it may wander away into the meadows and uplands to vary its diet with frogs and small mammals. Even more familiar is the little green heron or poke (*Ardea virescens*), which also is seen widely over the country. The night-herons, as their name indicates, stalk their prey by night, and during the day roost in companies—a characteristic common to most herons. The bitterns or stake-drivers are at home in reedy swamps, where they live singly or in pairs, and throughout the night, during times of migration, utter a booming noise resembling the driving of a stake into boggy ground. As a rule, the herons breed as they roost—in companies—building bulky platforms, usually in trees. The bitterns, on the other hand, secrete their nests on the ground in the rushes of their marshy home.

208. Cranes, rails, and coots (Paludicolæ).—In their external form the cranes and rails resemble the herons, but in their internal organization they differ considerably. They likewise inhabit marshy lands, but usually avoid wading, picking up the frogs, fish, and insects or plants along the shore or from the surface of the water. The cranes are comparatively rare in this country, yet one may occasionally meet with the whooping-cranes (*Grus americana*) and sand-hill cranes (*Grus mexicana*), especially in the South and West. They are said to mate for life, and annually repair to the same breeding-grounds, where they build their

nests of grass and weeds on the ground in marshy places.
The rails are more abundant, though rarely seen on account of their habit of skulking through the swamp
grasses. Only rarely do they take to the wing, and then
fly but a short distance, with their legs dangling awk-
wardly. Closely related to them are the coots or mud-hens
(*Fulica americana*), which may be distinguished, however,
by their slaty color, white bills, and lobed webs on the toes,
and consequent ability to swim. All over the United
States they may be seen resting on the shores of lakes or
quiet streams, or swimming on the surface gathering food.
The nest consists of a mass of floating reeds, which the
young abandon almost as soon as hatched.

209. **The snipes, sandpipers, and plovers (Limicolæ).**—The
snipes, sandpipers, and plovers are usually small birds,
widely scattered throughout the country wherever there
are sandy shores and marshes. In most species the legs
are long, and in connection with the slender, sensitive bill
fit the bird for picking up small animals in shallow water
or probing for them deep in the mud. During the greater
part of the year they travel in flocks, but at the nesting-
season disperse in pairs and build their nests in shallow depressions in the earth. The eggs are usually
streaked and spotted, in harmony with their surroundings,
as are the young, which leave the nest almost as soon as
hatched.

Fully fifty species of these shore-birds live within the
confines of the United States. Among these the woodcock
(*Philohela minor*) and snipe (*Gallinago delicata*) are abundant in many places inland, where they probe the moist soil
for food, and in turn are eagerly sought by the sportsman.
Even more familiar are the sandpipers and plovers, which
are especially common along the seacoast, and are also
abundantly represented by several species far inshore.
Among the latter are the well-known spotted sandpiper or
"tip-up" (*Actitis macularia*) and the killdeer plover (*Ægi-*

alitis vocifera), which inhabit the shores of lakes and
streams throughout the country.

210. **Quail, pheasants, grouse, and turkeys (Gallinæ).**—The
quail, grouse, and our domestic fowls are all essentially

Fig. 125.—California quail (*Lophortyx californicus*). Two-thirds natural size.

ground-birds, and their structure well adapts them to such
a life. The body is thick-set, the head small, and the beak
heavy for picking open and crushing the seeds and berries

upon which they live. The legs and feet are stout, and fitted for scratching or for running through grass and underbrush. Protective colors also prevent detection, but if close pressed they rise into the air with a rapid whirring of their stubby wings, and after a short flight settle to the ground again. During the breeding-season the male usually mates with a number of hens, which build rough nests in hollows in the ground, where they lay numerous eggs. The young are precocial.

The quail or bob-white (*Colinus virginianus*) and the ruffed grouse (*Bonasa umbellus*) occur throughout the Eastern States. Over the same area the wild turkey (*Meleagris gallopavo*) once extended, but is now almost extinct. The prairies of the middle West support the prairie-hen (*Tympanuchus americanus*), and the valleys and mountains of the far West are the home of several species of quails, some of which are beautifully crested.

211. **Pigeons and doves (Columbæ).**—The pigeons and doves belong to a small yet well-defined order, with upward of a dozen representatives in the United States. They are of medium size, with small head, short neck and legs, and among other distinguishing characters frequently possess a swollen, fleshy pad in which the nostrils are placed. In former years the passenger-pigeon (*Ectopistes migratorius*), inhabiting eastern North America, was probably the most common species in this country. Their flocks contained thousands, at times millions, of individuals, which often traveled hundreds of miles a day in search of food, to return at night to definite roosts—a trait which enabled the hunter to practically exterminate them. At present the mourning- or turtle-dove (*Zenaidura macroura*) is the most familiar and wide-spread of the wild forms. The domestic pigeons are all descendants of the common rock-dove (*Columba livia*) of Europe, the numerous varieties such as the tumblers, fantails, pouters, etc., being the product of man's careful selection. In the construction of the nest, usually

a rude platform of twigs, and in the care of the young both parents have a share. The young at hatching are blind, naked, and perfectly helpless, and are fed masticated food from the crops of the parents until able to subsist on fruits and seeds.

212. **Eagles, hawks, owls, etc. (Raptores).**—The birds of prey, all of which belong to this order, are carnivorous, often of large size and great strength, and are widely distributed throughout this country. The vultures live on carrion, some of the small hawks and owls on insects, while the majority capture small birds and mammals by the aid of powerful talons. In every case the beak is hooked, and the perfection of the organs of sight and hearing is unequaled by any other animal, man included. They live in pairs, and in many species mate for life. As a rule, the female incubates the eggs, and the male assists in collecting food.

Among the vultures, the turkey-buzzard (*Cathartes aura*) is most abundant throughout the United States, especially in the warmer portions, where it plays an important part as a scavenger. Of the several species of hawks, the white-rumped marsh-hawk (*Circus hudsonius*), the red-tailed hawk (*Buteo borealis*), the red-shouldered hawk (*Buteo lineatus*), and above all the bold though diminutive sparrow-hawk (*Falco sparverius*) are the most abundant and familiar. In the more unsettled regions live the golden eagle (*Aquila chrysaetus*) and bald eagle (*Haliaetus leucocephalus*). The owls are nocturnal, and not so often seen as the other birds of prey, yet the handsome and fierce barn or monkey-faced owl (*Strix pratincola*), and the larger species, such as the great gray owl (*Scotiaptex cinereua*), and the beautiful snowy owl (*Nyctea nyctea*), are more or less common, and occasionally seen. Much more abundant is the little screech-owl (*Megascops asio*), and in the Western States the burrowing-owl (*Speotyto cunicularia*), which lives in the burrows of the ground-squirrels and prairie-dogs. Fiercest and strongest of the tribe is the great horned owl (*Bubo virginianus*).

FIG. 126.—Golden eagle (*Aquila chrysaëtus*).

213. Cuckoos and kingfishers (Coccyges).—Omitting the order of parrots (*Psittaci*), whose sole representative in this country is the almost exterminated Carolina parrakeet

(*Conurus carolinensis*), we next arrive at the cuckoos and kingfishers, which differ widely in their habits. The black- or yellow-billed cuckoos or rain-crows are shy, retiring birds, with drab plumage, and though seldom seen are often fairly abundant, and are of much service in destroying insects. Unlike their shiftless European relatives, which lay their eggs in the nests of others birds, they build their own airy homes in some bush or hedgerow, and raise their brood with tender care. The belted kingfisher (*Ceryle alcyon*) is also of a retiring disposition, and spends much of its time on some branch overlooking the water, occasionally varying the monotony by dashing after a fish, or flying with rattling cry to another locality. Their nests are built in holes in banks, and six or eight young are annually reared.

214. **The woodpeckers (Pici).**—The woodpeckers are widely distributed throughout the world, and are preeminently fitted for an arboreal life. The beak is stout for chiseling open the burrows of wood-boring insects, which are extracted by the long and greatly protrusible tongue. The feet, with two toes directed forward and two backward, are adapted for clinging, and the stiff feathers of the tail serve to support the bird when resting. Almost all are bright-colored, with red spots on the head, at least in the males, which may further attract their mates by beating a lively tattoo with their beaks on some dry limb. The glossy white eggs are laid in holes in trees, and both parents are said to share the duties of incubation and feeding the young. Among the more abundant and well-known species is the yellowhammer or flicker (*Colaptes auratus*), which extends throughout the United States. Somewhat less widely distributed is the red-headed woodpecker (*Melanerpes erythrocephalus*), and the small black-and-white downy woodpecker (*Dryobaies pubescens*). This is often called sapsucker, but incorrectly so, as, like all but one of our other woodpeckers it feeds on insects. The yellow-bellied wood-

pecker (*Sphyrapicus varius*) is a real sapsucker, living on the juices of trees. A close relative of the red-headed woodpecker, the California woodpecker (*Melanerpes formicivorus*), is renowned for its habit of boring holes in bark and inserting the acorns of the live oak. Subsequently the bird returns, and breaking open the acorns, devours the grubs which have infested them, and apparently eats the acorns also.

215. **Swifts, humming-birds, etc. (Macrochires).**—The birds of this order are rapid, skilful fliers, and their wings are very long and pointed. The feet, on the other hand, are

FIG. 127.—Night-hawk (*Chordeiles virginianus*) on nest. Photograph by H. K. JOB.

small, relatively feeble, and adapted for perching or clinging. Accordingly, the insects upon which they feed are taken during flight by means of their open beaks. The night-hawk (*Chordeiles virginianus*), roosting lengthwise on a branch by day, at nightfall takes to the wing, and high in the air pursues its food after the fashion of a swallow. In the same haunts throughout the United States the whip-

poorwill (*Antrostomus vociferus*) occurs, sleeping by day,
but active at night. Neither of these birds constructs nests,
but lays its streaked and mottled eggs directly on the
ground. The chimney-swifts (*Chætura pelagica*), swallow-
like in general form and habits, but very unlike the swallows

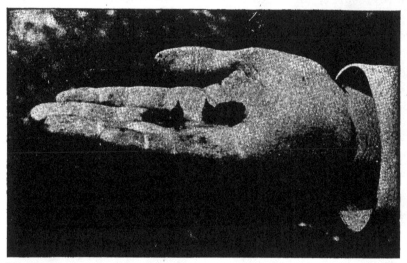

FIG. 128.—Anna hummers (one day old), showing short bill and small size of body.
Compare with last joint of little finger.

in structure, frequent hollow trees or unused chimneys, to
which they attach their shallow nests. The nearly related
humming-birds are chiefly natives of tropical America, only
a few species extending into the United States. Of these
the little, brilliantly colored, and pugnacious ruby throat
(*Trochilus colubris*) is the most widely distributed. Its
nest, like that of other hummers, is composed of moss and
lichens bound together with cobweb and lined with down.

216. **Perching birds** (**Passeres**).—The remaining birds,
over six thousand in all, belong to one order, the *Passeres*
or perchers. They are characterized by great activity,
interesting habits, frequently by exquisite powers of song,
and in addition to several other structural arrangements
have the feet adapted for perching. Their nesting habits

differ widely, but in every case the young are helpless at the time of hatching, and require the care of the parents.

The perchers constitute the greater number of the birds living in the meadows and woods, and are more or less

Fig. 129.—Anna hummer (*Calypte anna*) on nest.

common, and consequently familiar everywhere. Among the families into which the order is divided that of the fly-catchers (*Tyrannidæ*), the crows and jays (*Corvidæ*), the orioles and blackbirds (*Icteridæ*), the finches and sparrows (*Fringillidæ*), the swallows (*Hirundinidæ*), the warblers (*Mniotiltidæ*), the thrushes, robins, and bluebirds (*Turdidæ*), are the more familiar, though the others are equally interesting.

CHAPTER XVIII

THE MAMMALS

217. General characteristics.—The mammals, constituting the last and highest class of the vertebrates, comprise such forms as the opossum and kangaroo, the whales and porpoises, hoofed and clawed animals, the monkeys and man. All are warm-blooded, air-breathing animals, having the skin more or less hairy. The young are born alive, except in the very lowest forms, which lay eggs like reptiles, and for some time after birth are nourished by milk supplied from the mammary glands (hence the word *mammals*) of the mother. The skeleton is firm, the skull and brain within are relatively large, and, with few exceptions, four limbs are present.

Most of the mammals inhabit dry land. A number, however, such as the whales and seals, are aquatic; while others, such as the beavers, muskrats, etc., though not especially adapted for an aquatic life, are, nevertheless, active swimmers, and spend much of their time in the water.

Mammals tend to associate in companies, as we may witness among the ground-squirrels, prairie-dogs, rats, mice, and the seals and whales. In many cases they band for mutual protection, and often fight desperately for one another. Claws, hoofs, and nails are efficient weapons, and spiny hairs, as on the porcupines, bony plates, such as encircle the bodies of the armadillos, and thick skin and hair, serve as a protection. The hair is also frequently colored to harmonize the animal with its surroundings.

232

Some rabbits and hares in the far north don a white coat in the winter season.

218. **Skeleton.**—As in other vertebrates, the external form of mammals is dependent in large measure upon the internal skeleton. This consists of relatively compact bones, the cavities of which are filled with marrow. Those forming the skull are firmly united, and, as in other vertebrates, afford lodgment for several organs of special sense and for the brain, which, like that of the birds, completely fills the cavity in which it rests. The vertebral column to which the skull is attached differs considerably in length, but it invariably gives attachment to the ribs, and to the basal girdles supporting one or two pairs of limbs. Generally speaking, the number of bones in the head and trunk of all mammals is the same, so the variations we note in the species about us, for example, are simply due to differences of shape and proportion. As we are aware, there is a great dissimilarity between the length of the neck of man and that of the giraffe, yet the number of bones in each is precisely the same. On the other hand, the variations occurring in the limbs are often due to the actual disappearance of parts of the skeleton. Five digits in hand and foot is the rule, and yet, as we well know, the horse walks on the tip of its middle finger and toe, the others being represented by small, very rudimentary, splint bones attached far up the leg. The even-hoofed animals walk on two digits, two smaller hoofed toes being often plainly visible a short distance up the leg, as in the pig. In the whales the hind limbs have completely disappeared, and in the seals, where the fore limbs are modified, as in the whales, into flippers, the hind limbs show many signs of degeneration.

219. **Digestive system.**—Some mammals, such as man, monkeys, and pigs, are omnivorous; others, like the cud-chewers and gnawers, are vegetarians; and still others, like the foxes, weasels, and bears, are carnivorous. In

every case the food substances are acted on by a digestive system constructed on the same general plan as that in man, yet modified according to the specific work it is required to perform. The teeth especially afford a valuable indication of the animal's feeding habits, and, as we may notice later, are also of much value in classification. They consist of incisors used in biting, canines for tearing, and premolars and molars for crushing and grinding.

The remaining portions of the digestive tract, esophagus, stomach, and intestine, with their appended glands, are usually not unlike those possessed by the squirrel. The chief differences are in the size of the various regions. The stomach, for example, may be long and slender or of great dimensions, and its surface may further be increased by several lobes, which are especially well developed in the ruminants or cud-chewers. The intestine, relatively longer in the mammals than in any other class of vertebrates, also exhibits great differences in length and size. In the flesh-eating species its length is about three or four times the length of the body, while in the ruminants it is ten or twelve times the length of the animal.

220. **Nervous system and sense-organs.**—As before noted, the nervous system of mammals is characterized by the large size and great complexity of the brain. Even in the simpler species the cerebral hemispheres (large front lobes or the brain) are well developed, and in the higher forms of the ascending series they form by far the larger part of the brain. The sense-organs also are highly developed, and are constructed and located much as they are in man. The greatest variations occur in the eyes. In some of the burrowing animals they are usually small, and in some of the moles and mice may even be buried beneath the skin and very rudimentary. On the other hand, they are large and highly organized in nocturnal animals; more so, usually, than in those which hunt their prey by day. The ears also have different grades of perfection, which

appear to be correlated with the habits of the animal. Among the species of subterranean habits the sense of hearing is largely deficient; but, on the other hand, it is exceedingly keen in the ruminants, and enables them to detect their enemies at surprisingly great distances. In these creatures the outer ears are of large size and great mobility, and, placed as they are on the top of the head, serve to concentrate the sound-waves on the delicate apparatus within. In the mammals the sense of smell reaches its highest development, especially among the carnivores which scent their prey. On the other hand, it is said to be absent in the whales and very deficient in the seals. The sense of taste, closely related to that of smell, is located in taste-buds on the tongue, and is also more acute than in any other class of animals. The sense of touch, located over the surface of the body, is especially delicate on the tips of the fingers, the tongue, and lips, which often bear long tactile hairs, called whiskers or *vibrissæ*.

221. **Mental qualities.**—Correlated with the high degree of perfection of the brain and sense-organs the mammals show a higher degree of development of the intellectual faculties than any other class of animals. In many cases their acts are instinctive, and not the result of previous training and experience. Just as the duck hatched in an incubator instinctively takes to the water and pecks at its food, or as the bee builds its symmetrical comb, many of the mammals perform their duties day by day. On the other hand many other mammals are also undoubtedly intelligent. They possess the faculty of memory; they form ideas and draw conclusions; they exhibit anger, hatred, and self-sacrificing devotion for their companions and offspring that is different from that in man only in degree and not in kind. In fact, intelligence differs from instinct primarily in its power of choice among lines of action.

222. **Classification.**—Of the eleven orders into which the mammals have been divided eight are represented in this

Fig. 130.—Three-toed sloth (*Bradypus*).   About one-tenth natural size.

country. Of the other three the first (*Monotremes*) and
simplest of the eleven is represented by the duck-mole

Fig. 131.—Australian duck-mole (*Ornithorhynchus paradoxus*). One-fifth natural
size.

(*Ornithorhynchus*) living in the Australian rivers. Its
general appearance and mode of life are illustrated in

FIG. 132.—The manatee, or sea-cow (*Trichechus latirostris*). A living species of sea-cow related to the now extinct Steller's sea-cow.

Fig. 131. The duck-moles are the only mammals which lay eggs. These are deposited in a carefully constructed nest where the young are hatched. Another order (*Edentata*) includes a number of South and Central American forms, among which are the ant-eaters, armadillos, and tree-inhabiting sloths (Fig. 130). Still another order (*Sirenia*) includes the fish-shaped marine dugong and sea-cows or manatees (Fig. 132), of which one species is found occasionally on the Florida coast. The remaining orders are described in the succeeding sections.

223. **The opossums and** kangaroos (**Marsupialia**).—The lowest order of mammals represented in the United States

Fig. 133.—Opossum (*Didelphys virginiana*). One-tenth natural size. Photograph by W. H. Fisher.

is that of the marsupials. It includes the opossums and kangaroos, together with a number of comparatively small and unfamiliar animals living chiefly in and about Australia.

The opossums, fairly abundant throughout the warmer portions of this country, are rat-like creatures, with scaly tails, yellowish-white fur, large head, and pointed snout. Except at the breeding season they lead solitary lives, sleeping in the holes of trees by day and at night feeding on roots, birds, and fruits.

The kangaroos, familiar from specimens in menageries or museums, chiefly inhabit the plains of Australia. The giant gray kangaroos (*Macropus giganteus*), attaining a height of over six feet, go in herds, and owing to the great development of their hind limbs and tails are able, when alarmed, to travel with the swiftness of a horse. Several smaller species, some no larger than rabbits, live among the brush, and like their larger relatives crop the grass and tender herbage with sharp incisor teeth.

While the marsupials do not lay eggs as does the duck-mole, they allow them to develop within the body for a very short time only. Hence the young, when born, are scarcely more than an inch in length, and are blind, naked, and perfectly helpless. At once they are placed by the mother in the pouch of skin, or *marsupium*, on the under side of her body. In this the young are suckled and pro-tected until able to gather their own food and fight their own way.

224. Rodents or gnawers (Glires).—The rodents are a large group of mammals, including such forms as the rats, mice, squirrels, gophers, and rabbits. They are readily dis-tinguished by their clawed feet adapted for climbing or burrowing, and by large curved incisor teeth. Unlike ordinary teeth, they grow continually, and, owing to the restriction of the hard enamel to their front surfaces, wear away behind faster than in front, thus producing a chisel-like cutting edge.

The largest of our native rodents is the porcupine (*Erethizon dorsatus*), which ranges from Maine to Mexico, and attains a length of nearly three feet. Many of the hairs of the body have the form of stiff, barbed spines (Fig. 134), readily dislodged so that the animal requires no other wea. pon of defense. The rabbits and hares are of smaller size, and the cottontails especially are widely distributed. West of the Mississippi the jack-rabbits are familiar, and are

FIG. 134. —Porcupine (*Hystrix cristata*).   One-tenth natural size.—After BREHM.

famous for their great speed. Like the porcupines, they feed on leaves and grass, and are often very destructive. The mice, especially the field and white-footed mice, are abundant in woodland and meadow throughout the United States. The house-mouse (*Mus musculus*) is a native of Europe, as is the common rat (*M. decumanus*), which was imported over a century ago. The wood-rat (*Neotoma*), however, is native, and may be found in many localities from east to west. The muskrat (*Fiber zibethicus*), beaver (*Castor canadensis*), and woodchuck (*Arctomys monax*) were also more or less plentiful formerly, but in many localities are well-nigh exterminated. The squirrels, on the other hand, continue to exist in large numbers. The prairie-dogs, ground-squirrels, and chipmunks of the terrestrial species are of frequent occurrence, and of the tree-dwellers the fox, gray and red squirrels are well known in many sections of the United States.

225. **Insect-eating mammals (Insectivora).**—The shrews and moles belonging to this order are representatives of a large group of small animals, which, unlike the major number of rodents, live on insects. The shrews, of which there are several species in this country, are small, mouse-like creatures, nocturnal in their habits, and hence rarely seen. The moles are of much larger size, and owing to their burrowing proclivities scarcely ever appear above ground, but excavate elaborate burrows with their shovel-like feet, devouring the insects which fall in their way. The common mole (*Scalops aquaticus*) extends from the eastern seaboard to the Mississippi River, where it is replaced by the prairie-mole (*S. argenteus*), which extends far to the west, into a country inhabited by other species.

226. **The bats (Cheiroptera).**—The bats are also insectivorous, but their habits are widely different from those of the shrews and moles. The forearm and the fingers of the fore limbs are greatly elongated, and are connected by a thin papery membrane, which also includes the hind limbs and tail, and serves as an efficient organ of flight. During the day they remain suspended head downward in some dark cranny, awakening at nightfall to capture flying insects. Several species are found in this country, the most common being the little brown bat (*Vespertilio fuscus*), with small, fox-like face, large erect ears, and short olive-brown hair. The red bat (*Lasiurus borealis*) is also plentiful everywhere throughout the United States, and is distinguished from the preceding by its somewhat larger size and long reddish-brown fur.

227. **The whales and porpoises (Cete).**—The animals belonging to this order, the whales (Fig. 135), porpoises, and dolphins, are aquatic animals bearing a resemblance to fishes

only in external form.   The cylindrical body has no distinct
neck, the comparatively large head uniting directly with

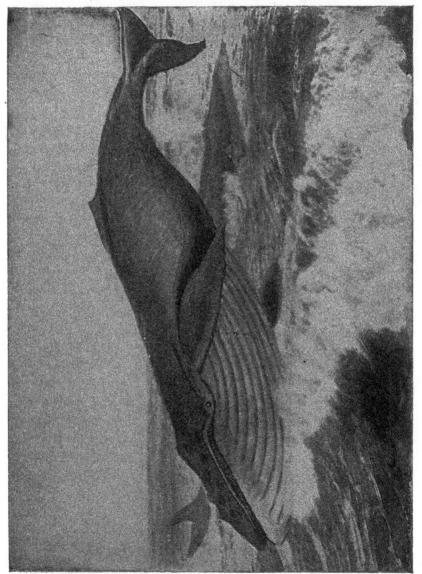

Fig. 135.—Humpback whale (*Megaptera versabilis*).   Attains a length of seventy feet.

the cylindrical body, which terminates in the tail with horizontally placed fins.   No external signs of hind limbs exist,

while the fore limbs are short and capable of being moved only as a whole. External ears are also absent. The eyes are exceedingly small, those of individuals attaining a length of from fifty to eighty feet, being in some species, at least, but little larger than those of an ox. These are often placed at the corners of the mouth. The nasal openings, often known as blow-holes, are situated on the forehead, and as the whale comes to the surface for air afford an outlet for the stream of breath and vapor often blown high in the air—a process known as spouting. In some of the whales, such as the dolphin, porpoise, and sperm-whales, the teeth persist throughout life, but in most of the larger species they never " cut " the gum, but early disappear, and their place is taken by large numbers of whalebone plates with frayed edges which act as strainers. The smaller-toothed forms (porpoises, dolphins, and several species of grampus) are frequently seen close to the shore, where they are usually actively engaged in capturing fish. On the other hand, the larger species, such as the humpback, right whale, and sulfurbottom, not uncommon along our coasts, especially to the northward, live on much smaller organisms. With open mouth these whales swim through the water until they collect a sufficient quantity of jelly-fishes, snails, and crustacea, then closing the mouth strain out the water through the whalebone fringes and swallow the residue.

As noted above, the animals of this order are almost wholly devoid of hair, but the heat of the body is retained by a thick layer of fat beneath the skin. This " blubber " also gives lightness to the body (as do the voluminous lungs), and, furthermore, yields large quantities of oil, which in former times made " whale-fishing " a profitable industry. The whales bear one, rarely two offspring, which are solicitously attended by the mother for a long time. The smaller species grow to a length of from five or eight feet (porpoises, dolphins) to twice this size (grampuses); while the larger whales, by far the largest of animals, range from thirty to over a hundred feet in length with a weight of many tons.

228. Hoofed mammals (Ungulata).—The order of hoofed animals or ungulates includes a large number of forms like the zebra, elephant, hippopotamus, giraffe, deer, and several other wild species, some of which are domesticated, such as horses, sheep, goats, and cattle. All of these animals walk on the tips of their toes, and the claws have become developed into hoofs. The order is divided into the odd-toed forms (perissodactyls), such as the rhinoceros with three toes and the horse with one, and the even-toed (artiodactyls), as the pigs with four, and the ox, deer, etc., with two toes. The even-toed forms are again divided into those which chew the cud (ruminants) and those which do not (non-ruminants). No living native odd-toed mammal exists in this country, and of the wild even-toed species all are ruminants. In the members of this latter group the swallowed food passes into a capacious sac (the paunch), is thoroughly moistened, and passed into the second division (the honeycomb), later to be regurgitated and ground by the powerful molars. It is then reswallowed, and undergoes successive treatment in the other two divisions of the stomach (the manyplies and reed) before entering the intestine.

Among the North American ruminants, the deer family (*Cervidæ*) is the best represented. In the more unsettled regions of the East the red deer is still common, and the same may be said of the white-tailed, black-tailed, and mule-deer of the West. Among the woods and lakes to the northward live the reindeer and caribou, and the largest of the deer family, the moose, which attains the size of the horse. Of nearly the same size is the wapiti or elk. In all of the above-mentioned species the horns, if present, are confined to the male (except in the reindeer), and are annually shed after the breeding season.

The native hollow-horned ruminants (*Bovidæ*) are at present confined to the Western plains, and comprise the pronghorn antelope (*Antilocapra americana*), the wary bighorn or Rocky Mountain sheep (*Ovis canadensis*), living in mountain fastnesses, and the buffalo or bison (*Bison americanus*). All of these species were formerly abundant, especially the pronghorn and buffalo, which roamed the plains by thousands, but their extermination has been nearly complete, small herds only persisting in a few wild, inaccessible regions, or protected in parks.

Our domestic sheep and cattle are probably the descendants of several wild species living in Europe and other portions of the world. Of the domesticated ungulates the horse is the direct descendant of Asiatic wild breeds; while the pig traces its ancestry back to the wild boar (*Sus scrofa*) of Europe, and probably a native species (*S. indicus*) of eastern Asia.

229. Flesh-eating mammals (Feræ).—The order of *Feræ* or Carnivora is typically exemplified by such animals as the lions, tigers, bears, dogs, cats, and seals, forms which differ from all other mammals by the large size of the canine teeth (often called dog-teeth) and the molars, which are adapted for cutting, not crushing. The limbs, terminated by four or five flexible digits, bear well-developed claws, which, together with the teeth, serve for tearing the prey. While the bears shuffle along on the soles of their feet, the greater number of species, as illustrated by the dog and cat, tread noiselessly on tiptoe. Almost all are fierce and bold, with remarkably keen senses and quick intelligence, and are the dreaded enemies of all other orders of mammals.

The largest land-inhabiting carnivora are the bears, of which the brown or cinnamon bear (*Ursus americanus*), inhabiting North America generally where not exterminated, and the huge grizzly (*Ursus horribilis*) of the Western mountains, are the best-known species. The former lives on berries and juicy herbs, while the grizzly prefers the flesh of animals which it kills. The raccoon (Fig. 136) (*Procyon lotor*) is found in wooded districts all over the United States, and its general appearance and thieving propensities are well known. Almost everything is accept-

FIG. 136.—Raccoon (*Procyon lotor*). Photograph by R. W. SHUFELDT.

able as an article of food, and its fondness for poultry and vegetables makes it an unmitigated nuisance. The otters, skunks, badgers, wolverenes, sables, minks, and weasels, while differing considerably in general appearance and habits, nevertheless belong to one family (the weasel family, *Mustelidæ*), and are more or less valued for their fur. Almost all are characterized by a fetid odor, especially the skunk, which is notoriously offensive, and in consequence is avoided by all other animals.

The dog family is represented by several widely distributed varieties of the red fox (*Vulpes pennsylvanicus*) and gray fox (*Urocyon cinereo-argentatus*), and by the coyotes

Fig. 137.—Silver fox (*Vulpes pennsylvanicus*, var. *argentatus*). Photograph by W. K. Fisher.

(*Canis latrans*) and wolves (*Canis nubilus*). The domestic dog (*Canis familiaris*) is probably the descendant of the wolf, and owing to man's careful breeding during thousands of years has formed several widely differing varieties.

The cat family, comprising the most powerful, savage, and keenest-scented carnivora, is represented by the lion, tiger, jaguar, and hyena. In this country the group is represented by the lynx (*Lynx canadensis*), the wildcat (*Lynx rufus*), and the panther or puma (*Felis concolor*), which attains the length of nearly five feet. The domestic cat has, like the dog, been domesticated for centuries, and has possibly descended from an African species (*Felis*

FIG. 138.—Panthers (*Felis concolor*) and peccaries (*Dicotyles torquatus*).

caffra), which was held sacred by the Egyptians, who embalmed them by thousands.

230. **Man-like mammals (Primates).**—The last and highest order of mammals, the Primates, includes the lemurs, monkeys, and man. The first of these are strange squirrel-like forms living chiefly in trees in Madagascar and neighboring regions where they feed on insects. The apes and monkeys are divided into Old and New World forms, which differ widely from each other. The American species are marked by flat noses, with the nostrils far apart. All are arboreal, many have long prehensile tails, and the digits bear nails, not claws. Among them are several species of marmosets, the howling monkeys (*Myocetes*), the spider-monkeys (*Ateles*), and the capuchins (*Cebus*), all of which are more or less common in captivity. In the Old World apes, on the other hand, the nostrils are close together and are directed downward, the tail is never prehensile, and in some cases is rudimentary, and may even disappear. The lowest species (the dog-like apes) include the large, clumsy baboons, among them the familiar blue-nosed mandrill (*Cynocephalus maimon*) and several other species of lighter frame, such as the long-tailed monkey (*Cercopithecus*) (Fig. 140), the tailless *Macacus*, common in menageries, and the Barbary ape, inhabiting northern Africa and extending over into Spain.

FIG. 139.—Baby orang-utan. From life.

The remaining anthropoid or man-like apes include the gibbons (*Hylobates*), orang-utan (*Simia*), gorilla (*Gorilla*), and chimpanzee (*Anthropo-*

FIG. 140.—A monkey (*Cercopithecus*) in a characteristic attitude of watchfulness.

pithecus). The gibbons, inhabiting southeastern Asia, possess arms of such length that they are able to touch their hands to the ground as they stand erect. They are thus adapted for a life in the trees, where they spend most of their time feeding on fruit, leaves, and insects. In the same district the orang occurs, walking when on the ground on its knuckles and the sides of its feet. It prefers the life in the trees, however, in which it builds nests serving for rest and concealment. The gorilla (Fig. 140), the largest of apes, attaining a height of over five feet and a weight of two hundred pounds, is a native of Africa, where it lives in families and subsists on fruits. The same region is the home of the chimpanzee, which in its various characteristics approaches most nearly to man.

Fig. 141.—Gorilla (*Gorilla*).

Man (*Homo sapiens*) is distinguished by the inability to oppose the big toe as he does his thumb—a feature associated with his erect position—and by the relatively enormous size of the brain. Even in an average four-year-old child or an Australian bushman the brain is twice as large as in the gorilla. With this relatively great development of the nervous system is correlated superior mental faculties, which together with social habits and powers of speech exalt man to a position far above the highest ape.

As usually understood, the family of man (*Hominidæ*) contains but a single species, cosmopolitan and highly variable. This species is "now split up into many subspecies or races, the native man of this continent, or 'American Indian,' being var. *americanus*. Other races now naturalized in America are: the Caucasian race, var. *europæus*; the Mongolian race, var. *asiaticus*; and the negro race, *afer*. The first of these is an immigrant from Europe, the second from Asia, and the third was brought hither from Africa by representatives of var. *europæus* to be used as slaves."

CHAPTER XIX

231. Birth, growth and development, and death.—Certain phenomena are familiar to us as occurring inevitably in the life of every animal. Each individual is born in an immature or young condition; it grows (that is, it increases in size), and develops (that is, changes more or less in structure), and dies. These phenomena occur in the succession of birth, growth and development, and death. But before any animal appears to us as an independent individual— that is, outside the body of the mother and outside of an egg (i. e., before birth or hatching, as we are accustomed to call such appearance)—it has already undergone a longer or shorter period of life. It has been a new living organism hours or days or months, perhaps, before its appearance to us. This period of life has been passed inside an egg, or as an egg or in the egg stage, as it is variously termed. The life of an animal as a distinct organism begins in an egg. And the true life cycle of an organism is its life from egg through birth, growth and development, and maturity to the time it produces new organisms in the condition of eggs. The life cycle is from egg to egg. Birth and growth, two of the phenomena readily apparent to us in the life of every animal, are two phenomena in the true life cycle. Death is a third inevitable phenomenon in the life of each individual, but it is not a part of the cycle. It is something outside.

232. Life cycle of simplest animals.—The simplest animals have no true egg stage, nor perhaps have they any true

254

death. The new *Amœbæ* are from their beginning like the full-grown *Amœba*, except as regards size. And the old *Amœba* does not die, because its whole body continues to live, although in two parts—the two new *Amœbæ*. The life cycle of the simplest animals includes birth (usually by simple fission of the body of the parent), growth, and some, but usually very little, development, and finally the reproduction of new individuals, not by the formation of eggs, but by direct division of the body.

233. **The egg.**—In our study of the multiplication of animals (Chapter VI) we learned that it is the almost univer-

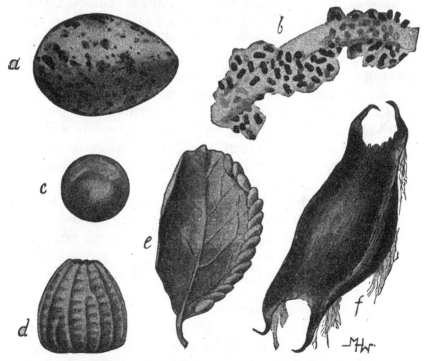

Fig. 142.—Eggs of different animals showing variety in external appearance. *a*, egg of bird ; *b*, eggs of toad ; *c*, egg of fish ; *d*, egg of butterfly ; *e*, eggs of katydid on leaf ; *f*, egg-case of skate.

sal rule among many-celled animals that each individual begins life as a single cell, which has been produced by the

fusion of two germ cells, a sperm cell from a male individual of the species and an egg cell from a female individual of the species. The single cell thus formed is called
the fertilized egg cell, and its subsequent development
results in the formation of a new individual of the same
species with its parents. Now, in the development of this
cell into a new animal, food is necessary, and sometimes a
certain amount of warmth. So with the fertilized egg cell
there is, in the case of all animals that lay eggs, a greater
or less amount of food matter—food yolk, it is called—gathered about the germ cell, and both germ cell and food yolk
are inclosed in a soft or hard wall. Thus is composed the
egg as we know it. The hen's egg is as large as it is because of the great amount of food yolk it contains. The
egg of a fish as large as a hen is much smaller than the
hen's egg; it contains less food yolk. Eggs (Fig. 142) may
vary also in their external appearance, because of the different kinds of membrane or shells which may inclose and
protect them. Thus the frog's eggs are inclosed in a thin
membrane and imbedded in a soft, jelly-like substance;
the skate's egg has a tough, dark-brown leathery inclosing
wall; the spiral egg of the bull-head sharks is leathery and
colored like the dark-olive seaweeds among which it lies;
and a bird's egg has a hard shell of carbonate of lime. But
in each case there is the essential fertilized germ cell; in
this the eggs of hen and fish and butterfly and cray-fish and
worm are alike, however much they may differ in size and
external appearance.

234. **Embryonic and post-embryonic development.**—Some
animals do not lay eggs, that is they do not deposit the fertilized egg cell outside of the body, but allow the development of the new individual to go on inside the body of the
mother for a longer or shorter period. The mammals and
some other animals have this habit. When such an animal issues from the body of the mother, it is said to be
born. When the developing animal issues from an egg

which has been deposited outside the body of the mother, it is said to hatch. The animal at birth or at time of hatching is not yet fully developed. Only part of its development or period of immaturity is passed within the egg or within the body of the mother. That part of its life thus passed within the egg or mother's body is called the embryonic life or embryonic stages of development; while that period of development or immaturity from the time of birth or hatching until maturity is reached is called the post-embryonic life or post-embryonic stages of development.

235. First stages in development.—The embryonic development is from the beginning up to a certain point practically identical for all many-celled animals—that is, there are cer-

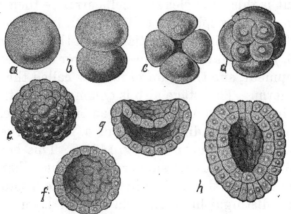

Fig. 143.—First stages in embryonic development of the pond snail (*Lymnæus*). *a*, egg cell; *b*, first cleavage; *c*, second cleavage; *d*, third cleavage; *e*, after numerous cleavages; *f*, blastula (in section); *g*, gastrula, just forming (in section); *h*, gastrula, completed (in section).—After RABL.

tain principal or constant characteristics of the beginning development which are present in the development of all many-celled animals. The first stage or phenomenon of development is the simple fission of the germ cell into halves (Fig. 143, *b*). These two daughter cells next divide so that there are four cells (Fig. 143, *c*); each of these divides, and this division is repeated until a greater or lesser number (varying with the various species or groups of animals) of cells is produced (Fig. 143, *d*). The phenomenon of repeated division of the germ cell, and usually the surrounding yolk, is called *cleavage*, and this cleavage is the first stage of development in the case of all many-celled animals. The first division of the germ cell produces usually two equal cells, but in some of the later divisions the new cells formed may not be equal. In some animals all the cleavage cells are of equal size; in some there are two sizes of cells. The germ or embryo animal consists now of a mass of few or many undifferentiated primitive cells lying together and usually forming a sphere (Fig. 143, *e*), or perhaps separated and scattered through the food yolk of the egg. The next stage of development is this: the cleavage cells arrange themselves so as to form a hollow sphere or ball, the cells lying side by side to form the outer circumferential wall of this hollow sphere (Fig. 143, *f*). This is called the *blastula* or *blastoderm* stage of development, and the embryo itself is called the blastula or blastoderm. This stage also is common to all the many-celled animals. The next stage in embryonic development is formed by the bending inward of a part of the blastoderm cell layer, as shown in Fig. 143, *g*. This bending in may produce a small depression or groove; but whatever the shape or extent of the sunken-in part of the blastoderm, it results in distinguishing the blastoderm layer into two parts, a sunken-in portion called the *endoblast* and the other unmodified portion called the *ectoblast*. *Endo-* means "within," and the cells of the endoblast often push so far into the original blastoderm cavity as to come into contact with the cells of the ectoblast and thus obliterate this cavity (Fig. 143, *h*). This third well-marked stage in the embryonic development is called the *gastrula* * stage, and it also

---

* This gastrula stage is not always formed by a bending in or invagination of the blastoderm, but in some animals is formed by the splitting off or delamination of cells from a definite limited region of

occurs in the development of all or nearly all many-celled animals.

236. **Continuity of development.**—In the case of a few of the simple many-celled animals the embryo hatches—that is, issues from the egg at the time of or very soon after reaching the gastrula stage. In the higher animals, however, development goes on within the egg or within the body of the mother until the embryo becomes a complex body, composed of many various tissues and organs. Al most all the development may take place within the egg,

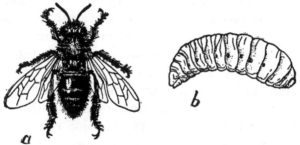

FIG. 144.—Honey-bee. *a.* adult worker; *b*, young or larval worker.

so that when the young animal hatches there is necessary little more than a rapid growth and increase of size to make it a fully developed, mature animal. This is the case with the birds: a chicken just hatched has most of the tissues and organs of a full-grown fowl, and is simply a little hen. But in the case of other animals the young hatches from the egg before it has reached such an advanced stage of development; a young star-fish or young crab or young honey-bee (Fig. 144) just hatched looks very different from its parent. It has yet a great deal of development to undergo before it reaches the structural condition of a fully developed and fully grown star-fish or crab or bee. Thus the development of some animals is almost

the blastoderm. Our knowledge of gastrulation and the gastrula stage is yet far from complete.

wholly embryonic development—that is, development within the egg or in the body of the mother—while the development of other animals is largely post-embryonic or larval
development, as it is often called.   There is no important
difference between embryonic and post-embryonic development. The development is continuous from egg-cell to
mature animal, and whether inside or outside of an egg it
goes on regularly and uninterruptedly.

237. **Development after the gastrula stage.**—The cells which
compose the embryo in the cleavage stage and blastoderm
stage, and even in the gastrula stage, are all similar; there
is little or no differentiation shown among them.   But from
the gastrula stage on development includes three important
things : the gradual differentiation of cells into various
kinds to form the various kinds of animal tissues; the
arrangement and grouping of these cells into organs and
body parts; and finally the developing of these organs
and body parts into the special condition characteristic of
the species of animal to which the developing individual
belongs.   From the primitive undifferentiated cells of the
blastoderm, development leads to the special cell types of
muscle tissue, of bone tissue, of nerve tissue; and from the
generalized condition of the embryo in its early stages development leads to the specialized condition of the body of
the adult animal.   Development is from the general to the
special, as was said years ago by the first great student of
development.

238. **Divergence of development.**—A star-fish, a beetle, a
dove, and a horse are all alike in their beginning—that is,
the body of each is composed of a single cell, a single structural unit.   And they are all alike, or very much alike,
through several stages of development; the body of each
is first a single cell, then a number of similar undifferentiated cells, and then a hollow sphere consisting of a single
layer of similar undifferentiated cells.   But soon in the
course of development the embryos begin to differ, and as

the young animals get further and further along in the course of their development, they become more and more different until each finally reaches its fully developed mature form, showing all the great structural differences between the star-fish and the dove, the beetle and the horse. That is, all animals begin development alike, but gradually diverge from each other during the course of development.

There are some extremely interesting and significant things about this divergence to which attention should be given. While all animals are alike structurally* at the beginning of development, so far as we can see, they do not all differ at the time of the first divergence in development. This first divergence is only to be noted between two kinds of animals which belong to different great groups or classes. But two animals of different kinds, both belonging to some one great group, do not show differences until later in their development. This can best be understood by an example. All the butterflies and beetles and grasshoppers and flies belong to the great group of animals called Insecta, or insects. There are many different kinds of insects, and these kinds can be arranged in subordinate groups, such as the Diptera, or flies, the Lepidoptera, or butterflies and moths, and so on. But all have certain structural characteristics in common, so that they are comprised in one great group or class—the Insecta. Another great group of animals is known as the Vertebrata, or back-boned animals. The class Vertebrata includes the fishes, the batrachians, the reptiles, the birds, and the mammals, each composing a subordinate group, but all characterized by the possession of a back-

---

* They are alike structurally, when we consider the cell as the unit of animal structure. That the egg cells of different animals may differ in their fine or ultimate structure, seems certain. For each one of these egg cells is destined to become some one kind of animal, and no other; each is, indeed, an individual in simplest, least developed condition of some one kind of animal, and we must believe that difference in kind of animals depends upon difference in structure in the egg itself.

bone, or, more accurately speaking, of a notochord, a back-bone-like structure. Now, an insect and a vertebrate diverge very soon in their development from each other; but two insects, such as a beetle and a honey-bee, or any two vertebrates, such as a frog and a pigeon, do not diverge from each other so soon. That is, all vertebrate animals diverge in one direction from the other great groups, but all the members of the great group keep together for some time longer. Then the subordinate groups of the Vertebrata, such as the fishes, the birds, and the others diverge, and still later the different kinds of animals in each of these groups diverge from each other. In the illustration (Fig. 145) on the opposite page will be seen pictures of the embryos of various vertebrate animals shown as they appear at different stages or times in the course of development. The embryos of a fish, a salamander, a tortoise, a bird, and a mammal, representing the five principal groups of the Vertebrata, are shown. In the upper row the embryos are in the earliest of all the stages figured, and they are very much alike. They show no obvious characteristics of fish or bird. Yet there are distinctive characteristics of the great class Vertebrata. Any of these embryos could readily be distinguished from an embryonic insect or worm or sea-urchin. In the second row there is beginning to be manifest a divergence among the different embryos, although it would still be a difficult matter to distinguish certainly which was the young fish and which the young salamander, or which the young tortoise and which the young bird. In the bottom row, showing the animals in a later stage of development, the divergence has proceeded so far that it is now plain which is a fish, which batrachian, which reptile, which bird, and which mammal.

239. **The laws or general facts of development.**—That the course of development of any animal from its beginning to fully developed adult form is fixed and certain is readily seen. Every rabbit develops in the same way; every grass-

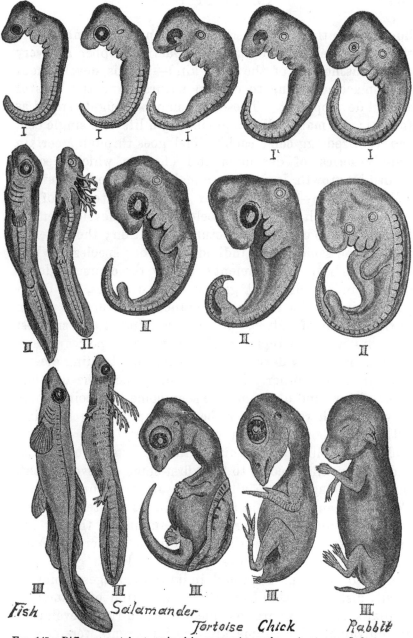

Fish   Salamander   Tortoise   Chick   Rabbit

FIG. 145.—Different vertebrate animal in successive embryonic stages. I, first or earliest of the stages figured; II, second of the stages; III, third or latest of the stages.—After HAECKEL.

hopper goes through the same developmental changes from single egg cell to the full-grown active hopper as every other grasshopper of the same kind—that is, development takes place according to certain natural laws, the laws of animal development. These laws may be roughly stated as follows : All many-celled animals begin life as a single cell, the fertilized egg cell ; each animal goes through a certain orderly series of developmental changes which, accompanied by growth, leads the animal to change from single cell to the many-celled, complex form characteristic of the species to which the animal belongs ; this development is from simple to complex structural condition ; the development is the same for all individuals of one species. While all animals begin development similarly, the course of development in the different groups soon diverges, the divergence being of the nature of a branching, like that shown in the growth of a tree. In the free tips of the smallest branches we have represented the various species of animals in their fully developed condition, all standing clearly apart from each other. But in tracing back the development of any kind of animal, we soon come to a point where it very much resembles or becomes apparently identical with some other kind of animal, and going further back we find it resembling other animals in their young condition, and so on until we come to that first stage of development, that trunk stage, where all animals are structurally alike. To be sure, any animal at any stage in its existence differs absolutely from any other kind of animal, in that it can develop into only its own kind of animal. There is something inherent in each developing animal that gives it an identity of its own. Although in its young stages it may be hardly distinguishable from some other kind of animal in similar stages, it is sure to come out, when fully developed, an individual of the same kind as its parents were or are. The young fish and the young salamander in the upper row in Fig. 145 seem very much alike, but one embryo is sure to develop into a fish and the other into a salamander. This certainty of an embryo to become an individual of a certain kind is called the law of heredity.

240. **The significance of the facts of development.**—The significance of the developmental phenomena is a matter about which naturalists have yet very much to learn. It is believed, however, by practically all naturalists that many of the various stages in the development of an animal correspond to or repeat the structural condition of the animal's ancestors. Naturalists believe that all backboned or vertebrate animals are related to each other through being descended from a common ancestor, the first or oldest backboned animal. In fact, it is because all these backboned animals—the fishes, the batrachians, the reptiles, the birds, and the mammals—have descended from a common ancestor that they all have a backbone. It is believed that the descendants of the first backboned animal have in the course of many generations branched off little by little from the original type until there came to exist very real and obvious differences among the backboned animals—differences which among the living backboned animals are familiar to all of us. The course of development of an individual animal is believed by many naturalists to be a very rapid, and evidently much condensed and changed, recapitulation of the history which the species or kind of animal to which the developing individual belongs has passed through in the course of its descent through a long series of gradually changing ancestors. If this is true, then we can readily understand why the fish and the salamander and tortoise and bird and rabbit are all so much alike in their earlier stages of development, and gradually come to differ more and more as they pass through later and later developmental stages.

Some naturalists believe that the ontogenetic stages are not as significant in throwing light upon the evolutionary history of the species as just indicated. Some think that when the earlier stages of one species correspond pretty closely with the early stages of another, we have a good basis for making up our minds about relationship between the two species. But it is certainly not obvious why we should have a similarity among the younger stages of different animals and no correspondence among the older stages of more recent animals with the younger stages of more ancient ones. But on the other hand it is certainly true that a too specific application of the broad generalization that ontogeny repeats phylogeny has led to numerous errors of interpreting genealogic relationship.

241. **Metamorphosis.** –While a young robin when it hatches from the egg or a young kitten at birth resembles its parents, a young star-fish or a young crab or a young butterfly when hatched does not at all resemble its parents. And while the young robin after hatching becomes a fully grown robin simply by growing larger and undergoing comparatively slight developmental changes, the young star-fish or young butterfly not only grows larger, but undergoes some very striking developmental changes; the body changes very much in appearance. Marked changes in the body of an animal during post-embryonic or larval development constitute what is called *metamorphic development*, or the animal is said to undergo or to show *metamorphosis* in its development. Metamorphosis is one of the most interesting features in the life history or development of animals, and it can be, at least as far as its external aspects are concerned, very readily observed and studied.

242. **Metamorphosis among insects.**—All the butterflies and moths show metamorphosis in their development. So do many other insects, as the ants, bees, and wasps, and all the flies and beetles. On the other hand, many insects do not show metamorphosis, but, like the birds, are hatched from the egg in a condition plainly resembling the parents. A grasshopper (Fig. 146) is a convenient example of an insect without metamorphosis, or rather, as there are, after all,

a few easily perceived changes in its post-embryonic development, of an insect with an "incomplete metamorphosis." The eggs of grasshoppers are laid in little packets of several score half an inch below the surface of the ground. When the young grasshopper hatches from the egg it is of course very small, but it is plainly recognizable as a grasshopper. But in one important character it differs from the adult, and that is in its lack of wings. The adult grasshopper has two pairs of wings; the just hatched young or larval grasshopper has no wings at all. The young grasshopper feeds voraciously and grows rapidly.

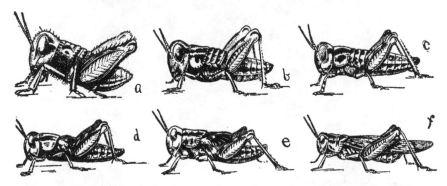

Fig. 146.—Post-embryonic development (incomplete metamorphosis) of the Rocky Mountain locust (*Melanoplus spretus*). *a, b, c, d, e,* and *f,* successive developmental stages from just hatched to adult individual.—After EMERTON.

In a few days it molts, or casts its outer skin (not the true skin, but a thin, firm covering or outer body wall composed of a substance called chitin, which is secreted by the cells of the true skin). In this second larval stage there can be seen the rudiments of four wings, in the condition of tiny wing pads on the back of the middle part of the body (the thorax). Soon the chitinous body covering is shed again, and after this molt the wing pads are markedly larger than before. Still another molt occurs, with another increase in size of the developing wings, and after a fifth and last molt the wings are fully developed, and the grasshopper is no longer in a larval or immature condition, but is full grown and adult.

For example of complete metamorphosis among insects we may choose a butterfly, the large red-brown butterfly

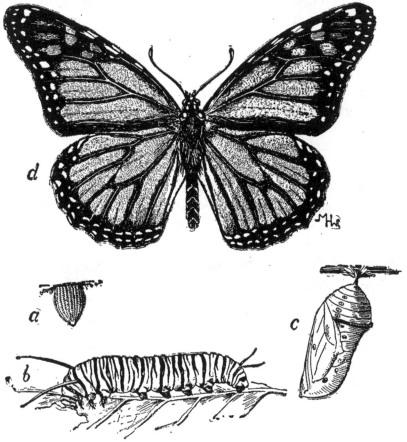

FIG. 147.—Metamorphosis of monarch butterfly (*Anosia plexippus*). *a*, egg ; *b*, larva ; *c*, pupa ; *d*, imago or adult.

common in the United States and called the monarch or milkweed butterfly (*Anosia plexippus*). The eggs (Fig. 147, *a*) of this butterfly are laid on the leaves of various kinds of milkweed (*Asclepias*). The larval butterfly or butterfly larva or caterpillar (as the first young stage of the butterflies and moths is usually called), which hatches from the egg in three or four days, is a creature bearing little or no resemblance to the beautiful winged adult. The larva is worm-like, and instead of having three pairs of legs like the butterfly it has eight pairs; it has biting jaws in its mouth with which it nips off bits of the green milkweed leaves, instead of having a long, slender, sucking proboscis for drinking flower nectar as the butterfly has.

The body of the crawling worm-like larva (Fig. 147, *b*) is greenish yellow in color, with broad rings or bands of shining black. It has no wings, of course. It eats voraciously, grows rapidly and molts. But after the molting there is no appearance of rudimentary wings; it is simply a larger wormlike larva. It continues to feed and grow, molting several times, until after the fourth molt it appears no longer as an active, crawling, feed-

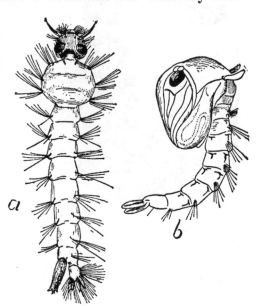

Fig. 148.—Metamorphosis of mosquito (*Culex*).
a, larva ; *b*, pupa.

ing, worm-like larva, but as a quiescent, non-feeding pupa or chrysalis (Fig. 147, *c*). The immature butterfly is now greatly contracted, and the outer chitinous wall is very thick and firm. It is bright green in color with golden dots. It is fastened by one end to a leaf of the milkweed, where it hangs immovable for from a few days to two weeks. Finally, the chitin wall of the chrysalis splits, and there issues the full-fledged, great, four-winged, red-brown butterfly (Fig. 147, *d*). Truly this is a metamorphosis, and a startling one. But we know that development in other animals is a gradual and continuous process, and so it is in the case of the butterfly. The gradual changing is masked by the outer covering of the body in both larva and pupa. It is only at each molting or throwing off of this unchanging, unyielding chitin armor that we perceive how far this change has gone. The longest time of concealment is that during the pupal or chrysalis stage, and the results of the changing or development when finally revealed by the splitting of the pupal case are hence the most striking.

Fig. 149.—Larva of a butterfly just changing into pupa (making last larval molt). Photograph from nature.

243. **Metamorphosis among other animals.**—Many other animals, besides insects and frogs and toads, undergo metamorphosis. The just-hatched sea-urchin does not resemble a fully developed sea-urchin at all. It is a minute wormlike creature, provided with cilia or vibratile hairs, by means of which it swims freely about. It changes next into a curious boot-jack shaped body called the pluteus stage (Fig. 150). In the pluteus a skeleton of lime is formed, and the final true sea-urchin body begins to appear inside the pluteus, developing and growing by using up the body substance of the pluteus. Star-fishes, which are closely related to sea-

urchins, show a similar metamorphosis, except that there is no pluteus stage, the true star-fish-shaped body forming, within and at the expense of the first larval stage, the ciliated free-swimming stage.

A young crab just issued from the egg (Fig. 151) is a very different appearing creature from the adult or fully developed crab. The body of the crab in its first larval stage is composed of a short, globular portion, furnished with conspicuous long spines and a relatively long, jointed tail.

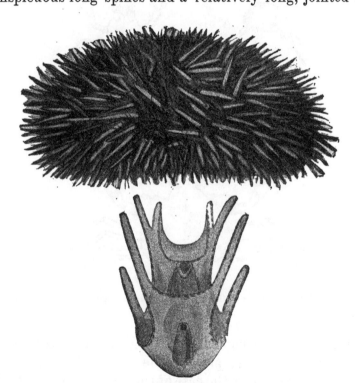

Fig. 150.—Metamorphosis of sea-urchin. Upper figure the adult, lower ngure the pluteus larva.

This is called the zoëa stage. The zoëa changes into a stage called the megalops, which has many characteristics of the adult crab condition, but differs especially from it in the possession of a long, segmented tail, and in having the front half of the body longer than wide. The crab in the megalops stage looks very much like a tiny lobster or shrimp. The tail soon disappears and the body widens, and the final stage is reached.

Interesting examples of metamorphosis occur in nearly all species of the animal kingdom, those mentioned being,

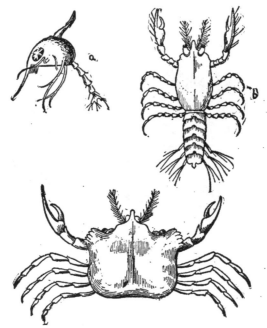

FIG. 151.—Metamorphosis of the crab. *a*, the zoëa stage ; *b*, the megalops ; *c*, the adult.

perhaps, the most conspicuous. In many families of fishes the changes which take place in the course of the life cycle are almost as great as in the case of the insect or the toad. In the lady-fish (*Albula vulpes*) the very young are ribbon-like in form, with small heads and very loose texture of the tissues, the body substance being jelly-like and transparent. As the fish grows older the body becomes more compact, and therefore shorter and slimmer. After shrinking to the texture of an ordinary fish, its growth in size begins normally, although

it has steadily increased in actual weight. Many herring, eels, and other soft-bodied fishes pass through stages similar to those seen in the lady-fish. Another type of development is illustrated in the sword-fish. The young has a bony head, bristling with spines. As it grows older the spines disappear, the skin grows smoother, and, finally, the bones of the upper jaw grow together, forming a prolonged sword, the teeth are lost and the fins become greatly modified. Fig. 152 shows three of these stages of growth. The

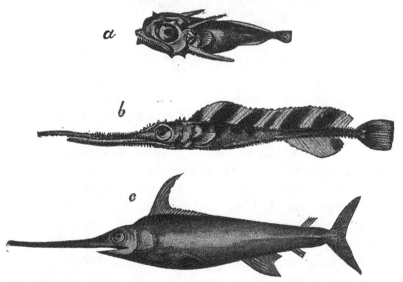

FIG. 152.—Three stages in the development of the sword-fish (*Xiphias gladius*). *a*, very young; *b*, older; *c*, adult.—Partly after LÜTKEN.

flounder or flat-fish (Fig. 152) when full grown lies flat on one side when swimming or when resting in the sand on the bottom of the sea. The eyes are both on the upper side of the body, and the lower side is blind and colorless. When the flounder is hatched it is a transparent fish, broad and flat, swimming vertically in the water, with an eye on each side. As its development (Fig. 153) goes on it rests itself obliquely on the bottom, the eye of the lower side turns upward, and as growth proceeds it passes gradually around the forehead, its socket moving with it, until both eyes and sockets are transferred by twisting of the skull to

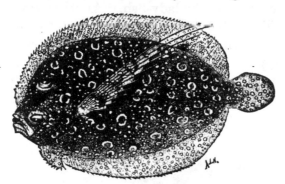

FIG. 153.—The wide eyed flounder (*Platophrys lunatus*). Adult, showing both eyes on upper side of head.

the upper side. In some related forms or soles the small eye passes through the head and not around it, appearing finally in the same socket with the other eye.

Thus in almost all the great groups of animals we find certain kinds which show metamorphosis in their post-embryonic development. But metamorphosis is simply development; its striking and extraordinary features are usually due to the fact that the orderly, gradual course of the development is revealed to us only occasionally, with the result of giving the impression that the development is proceeding by leaps and bounds from one strange stage to

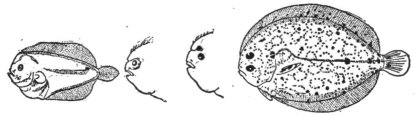

FIG. 154.—Development of a flounder (after EMERY). The eyes in the young flounder are arranged normally, one on each side of head.

another. If metamorphosis is carefully studied it loses its aspect of marvel, although never its great interest.

244. Duration of life.—After an animal has completed its development it has but one thing to do to complete its life cycle, and that is the production of offspring. When it has laid eggs or given birth to young, it has insured the beginning of a new life cycle. Does it now die? Is the business of its life accomplished? There are many animals which die immediately or very soon after laying eggs. The May-flies—ephemeral insects which issue as winged adults from ponds or lakes in which they have spent from one to three years as aquatic crawling or swimming larvæ, flutter about for an evening, mate, drop their packets of fertilized eggs into the water, and die before the sunrise — are extreme examples of the numerous kinds of animals whose adult life lasts only long enough for mating and egg-laying. But elephants live for two hundred years. Whales probably live longer. A horse lives about thirty years, and so may a cat or toad. A sea-anemone, which was kept in an aquarium, lived sixty-six years.

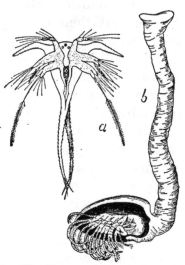

FIG. 155.—Metamorphosis of a barnacle (*Lepas*). *a*, larva ; *b*, adult.

Cray-fishes may live twenty years. A queen bee was kept in captivity for fifteen years. Most birds have long lives—the small song birds from eight to eighteen years, and the great eagles and vultures up to a hundred years or more. On the other hand, among all the thousands of species of insects, the individuals of very few indeed live more than a year; the adult life of most insects being but a few days or weeks, or at best months. Even among the higher animals, some are very short-lived. In Japan is a small fish (*Solaux*) which probably lives but a year, ascending the rivers in numbers when young in the spring, the whole mass of individuals dying in the fall after spawning.

Naturalists have sought to discover the reason for these extraordinary differences in the duration of life of different animals, and while it can not be said that the reason or reasons are wholly known, yet the probability is strong that the duration of life is closely connected with, or dependent upon, the conditions attending the production of offspring. It is not sufficient, as we have learned from our study of the multiplication of animals (Chapter VI), that an adult animal shall produce simply a single new individual of its kind, or even only a few. It must produce many, or if it produces comparatively few it must devote great care to the rearing of these few, if the perpetuation of the species is to be insured. Now, almost all long-lived animals are species which produce but few offspring at a time, and reproduce only at long intervals, while most short-lived animals produce a great many eggs, and these all at one time. Birds are long-lived animals; as we know, most of them lay eggs but once a year, and lay only a few eggs each time. Many of the sea birds which swarm in countless numbers on the rocky ocean islets and great sea cliffs lay only a single egg once each year. And these birds, the guillemots and murres and auks, are especially long-lived. Insects, on the contrary, usually produce many eggs, and all of them in a short time. The May-fly, with its one evening's lifetime, lets fall from its body two packets of eggs and then dies. Thus the shortening of the period of reproduction with the production of a great many offspring seem to be always associated with a short adult lifetime; while a long period of reproduction with the production of few offspring at a time and care of the offspring are associated with a long adult lifetime.

There seems also to be some relation between the size of animals and the length of life. As a general rule, large animals are long-lived and small animals have short lives.

245. The number of young.—There is great variation in the number of young produced by different species of animals. Among the animals we know familiarly, as the mammals, which give birth to young alive, and the birds, which lay eggs, it is the general rule that but few young are produced at a time, and the young are born or eggs are laid only once or perhaps a few times in a year. The robin lays five or six eggs once or twice a year; a cow may produce a calf each year. Rabbits and pigeons are more prolific, each having several broods a year. But when we observe the multiplication of some of the animals whose habits are not so familiar to us, we find that the production of so few young is the exceptional and not the usual habit. A lobster lays ten thousand eggs at a time; a queen bee lays about five million eggs in her life of four or five years. A female white ant, which after it is full grown does nothing but lie in a cell and lay eggs, produces eighty thousand eggs a day steadily for several months. A large codfish was found on dissection to contain about eight million eggs.

If we search for some reason for this great difference in fertility among different animals, we may find a promising clew by attending to the duration of life of animals, and to the amount of care for the young exercised by the parents. We find it to be the general rule that animals which live many years, and which take care of their young, produce but few young; while animals which live but a short time, and which do not care for their young, are very prolific. The codfish produces its millions of eggs; thousands are eaten by sculpins and other predatory fishes before they are hatched, and other thousands of the defenseless young fish are eaten long before attaining maturity. Of the great number produced by the parent, a few only reach maturity and produce new young. But the eggs of the robin are hatched and protected, and the helpless fledglings are fed and cared for until able to cope with their natural enemies. In the next year another brood is carefully reared, and so on for the few years of the robin's life.

Under normal conditions in any given locality the number of individuals of a certain species of animal remains about the same. The fish which produces tens of thousands of eggs and the bird which reproduces half a dozen eggs a year maintain equally well their numbers. In one case a few survive of many born; in the other many (relatively) survive of the few born; in both cases the species is effectively maintained. In general, no agency for the perpetuation of the species is so effective as that of care for the young.

246. **Death.**—At the end comes death. After the animal has completed its life cycle, after it has done its share toward insuring the perpetuation of its species, it dies. It may meet a violent death, may be killed by accident or by enemies, before the life cycle is completed. And this is the fate of the vast majority of animals which are born or hatched. Or death may come before the time for birth or hatching. Of the millions of eggs laid by a fish, each egg a new fish in simplest stage of development, how many or rather how few come to maturity, how few complete the cycle of life!

Of death we know the essential meaning. Life ceases and can never be renewed in the body of the dead animal. It is important that we include the words "can never be renewed," for to say simply that "life ceases," that is, that the performance of the life processes or functions ceases, is not really death. It is easy to distinguish in most cases between life and death, between a live animal and a dead one, yet there are cases of apparent death or a semblance of death which are very puzzling. The test of life is usually taken to be the performance of life functions, the assimilation of food and excretion of waste, the breathing in of oxygen, and breathing out of carbonic-acid gas, movement, feeling, etc. But some animals can actually suspend all of these functions, or at least reduce them to such a minimum that they can not be perceived by the strictest examination, and yet not be dead. That is, they can renew again the performance of the life processes. Bears and some other animals, among them many insects, spend the winter in a state of death-like sleep. Perhaps it is but sleep; and yet hibernating insects can be frozen solid and remain frozen for weeks and months, and still retain the power of actively living again in the following spring. Even more remarkable is the case of certain minute animals called *Rotatoria* and of others called *Tardigrada*, or bear-animalcules. These bear-animalcules live in water. If the water dries up, the animalcules dry up too; they shrivel up into formless little masses and become desiccated. They are thus simply dried-up bits of organic matter; they are organic dust. Now, if after a long time—years even—one of these organic dust particles, one of these dried-up bear-animalcules is put into water, a strange thing happens. The body swells and stretches out, the skin becomes smooth instead of all wrinkled and folded, and the legs appear in normal shape. The body is again as it was years before, and after a quarter of an hour to several hours (depending on the length of time the animal has lain dormant and dried) slow movements of the body parts begin, and soon the animalcule crawls about, begins again its life where it had been interrupted. Various other small animals, such as vinegar eels and certain Protozoa, show similar powers. Certainly here is an interesting problem in life and death.

When death comes to one of the animals with which we are familiar, we are accustomed to think of its coming to the whole body at some exact moment of time. As we stand beside a pet which has been fatally injured, we wait until suddenly we say, "It is dead." As a matter of fact, it is difficult to say when death occurs. Long after the heart ceases to beat, other organs of the body are alive—that is, are able to perform their special functions. The muscles can contract for minutes or hours (for a short time in warm-blooded, for a long time in cold-blooded animals) after the animal ceases to breathe and its heart to beat. Even longer live certain cells of the body, epecially the amœboid white blood-corpuscles. These cells, very like the *Amœba* in character, live for days after the animal is, as we say, dead. The cells which line the tracheal tube leading to the lungs bear cilia or fine hairs which they wave back and forth. They continue this movement for days after the heart has ceased beating. Among cold-blooded animals, like snakes and turtles, complete cessation of life functions comes very slowly, even after the body has been literally cut to pieces.

Thus it is essential in defining death to speak of a complete and permanent cessation of the performance of the life processes.

CHAPTER XX

247. **The crowd of animals.**—All animals feed upon living
organisms, or on their dead bodies. Hence each animal
throughout its life is busy with the destruction of other
organisms, or with their removal after death. If those
creatures upon which others feed are to hold their own, there
must be enough born or developed to make good the drain
upon their numbers. If the plants did not fill up their
ranks and make good their losses, the animals that feed
on them would perish. If the plant-eating animals were
destroyed, the flesh-eating animals would in turn disappear.
But, fortunately, there is a vast excess in the process of
reproduction. More plants sprout than can find room to
grow. More animals are born than can possibly survive.
The process of increase among animals is correctly spoken
of as multiplication. Each species tends to increase in
geometric ratio, but as it multiplies its members it finds
the world already crowded with other species doing the
same thing. A single pair of any species whatsoever, if not
restrained by adverse conditions, would soon increase to
such an extent as to fill the whole world with its progeny.
An annual plant producing two seeds only would have
1,048,576 descendants at the end of twenty-one years, if
each seed sprouted and matured. The ratio of increase is
therefore a matter of minor importance. It is the ratio of
net increase above loss which determines the fate of a species. Those species increase in numbers whose gain exceeds

281

the death rate, and those which "live beyond their means" must sooner or later disappear. One of the most abundant of birds is the fulmar petrel, which lays out one egg yearly. It has but few enemies, and this low rate of increase suffices to cover the seas within its range with petrels.

It is difficult to realize the inordinate numbers in which each species would exist were it not for the checks produced by the presence of other animals. Certain Protozoa at their normal rate of increase, if none were devoured or destroyed, might fill the entire ocean in about a week. The conger-eel lays, it is said, 15,000,000 eggs. If each egg grew up to maturity and reproduced itself in the same way in less than ten years the sea would be solidly full of conger-eels. If the eggs of a common house-fly should develop, and each of its progeny should find the food and temperature it needed, with no loss and no destruction, the people of a city in which this might happen could not get away soon enough to escape suffocation from a plague of flies. Whenever any insect is able to develop a large percentage of the eggs laid, it becomes at once a plague. Thus originate plagues of grasshoppers, locusts, and caterpillars. But the crowd of life is such that no great danger exists. The scavenger destroys the decaying flesh where the fly would lay its eggs. Minute creatures, insects, bacteria, Protozoa are parasitic within the larva and kill it. Millions of flies perish for want of food. Millions more are destroyed by insectivorous birds, and millions are slain by parasites. The final result is that from year to year the number of flies does not increase. Linnæus once said that "three flies would devour a dead horse as quickly as a lion." Equally soon would it be devoured by three bacteria, for the decay of the horse is due to the decomposition of its flesh by these microscopic plants which feed upon it. "Even slow-breeding man," says Darwin, "has doubled in twenty-five years. At this rate in less than a thousand years there would literally not be standing room for his progeny. The elephant is reckoned the slowest breeder of all known animals. It begins breeding when thirty years old and goes on breeding until ninety years old, bringing forth six young in the interval, and surviving till a hundred years old. If this be so, after about eight hundred years there would be 19,000,000 elephants alive, descended from the first pair." A few years more of the unchecked multiplication of the elephant and every foot of land on the earth would be covered by them.

Yet the number of elephants does not increase. In general, the numbers of every species of animal in the state of Nature remain about stationary. Under the influence of man most of them slowly diminish. There are about as many squirrels in the forest one year as another, about as many butterflies in the field, about as many frogs in the pond. Wolves, bears, deer, wild ducks, singing birds, fishes, tend to grow fewer and fewer in inhabited regions, because the losses from the hand of man are added to the losses in the state of Nature.

It has been shown that at the normal rate in increase of English sparrows, if none were to die save of old age, it would take but twenty years to give one sparrow to every square inch in the State of Indiana. Such an increase is actually impossible, for more than a hundred other species of similar birds are disputing the same territory with the power of increase at a similar rate. There can not be food and space for all. With such conditions a struggle is set up between sparrow and sparrow, between sparrow and other birds, and between sparrow and the conditions of life. Such a conflict is known as the struggle for existence.

248. **The struggle for existence.**—The struggle for existence is threefold: (*a*) among individuals of one species, as sparrow and sparrow; (*b*) between individuals of different species, as sparrow with bluebird or robin; and (*c*) with the conditions of life, as the effort of the sparrow to keep warm in winter and to find water in summer. All three forms of this struggle are constantly operative and with

every species. In some regions the one phase may be more destructive, in others another. Where the conditions of life are most easy, as in the tropics, the struggle of species with species, of individual with individual, is the most severe.

No living being can escape from any of these three phases of the struggle for existence. For reasons which we shall see later, it is not well that any should escape, for " the sheltered life," the life withdrawn from the stress of effort, brings the tendency to degeneration.

Because of the destruction resulting from the struggle for existence, more of every species are born than can possibly find space or food to mature. The majority fail to reach their full growth because, for one reason or another, they can not do so. All live who can. Each strives to feed itself, to save its own life, to protect its young. But with all their efforts only a portion of each species succeed.

249. **Selection by Nature.**—But the destruction in Nature is not indiscriminate. In the long run those least fitted to resist attack are the first to perish. It is the slowest animal which is soonest overtaken by those which feed upon it. It is the weakest which is crowded away from the feeding-place by its associates. It is the least adapted which is first destroyed by extremes of heat and cold. Just as a farmer improves his herd of cattle by destroying his weakest or roughest calves, reserving the strong and fit for parentage, so, on an inconceivably large scale, the forces of Nature are at work purifying, strengthening, and 'fitting to their surroundings the various species of animals. This process has been called natural selection, or the survival of the fittest. But by fittest in this sense we mean only best adapted to the surroundings, for this process, like others in Nature, has itself no necessarily moral element. The song-bird becomes through this process more fit for the song-bird life, the hawk becomes more capable of killing and tear-

ing, and the woodpecker better fitted to extract grubs from the tree.

In the struggle of species with species one may gain a little one year and another the next, the numbers of each species fluctuating a little with varying circumstances, but after a time, unless disturbed by the hand of man, a point will be reached when the loss will almost exactly balance the increase. This produces a condition of apparent equilibrium. The equilibrium is broken when any individual or group of individuals becomes capable of doing something more than hold its own in the struggle for existence.

When the conditions of life become adverse to the existence of a species it has three alternatives, or, better, one of three things happens, namely, migration, adaptation, extinction. The migration of birds and some other animals is a systematic changing of environment when conditions are unfavorable to life. When the snow and ice come, the fur-seal forsakes the islands on which it breeds, and which are its real home, and spends the rest of the year in the open sea, returning at the close of winter. Some other animals migrate irregularly, removing from place to place as conditions become severe or undesirable. The Rocky Mountain locusts, which breed on the great plateau along the eastern base of the Rocky Mountains, sometimes increase so rapidly in numbers that they can not find enough food in the scanty vegetation of this region. Then great hosts of them fly high into the air until they meet an air current moving toward the southeast. The locusts are borne by this current or wind hundreds of miles, until, when they come to the great grain-growing Mississippi Valley, they descend and feed to their hearts' content, and to the dismay of the Nebraska and Kansas farmer. These great forced migrations used to occur only too often, but none has taken place since 1878, and it is probable that none will ever occur again. With the settlement of the Rocky Mountain plateau by farmers, food is plenty at home. And the constant fighting of the locusts by the farmers, by plowing up their eggs, and crushing and burning the young hoppers, keeps down their numbers.

Another animal of interesting migratory habits is the lemming, a mouse-like animal nearly as large as a rat, which lives in the arctic regions. At intervals varying from five to twenty years the cultivated lands of Norway and Sweden, where the lemming is ordinarily unknown, are overrun by vast numbers of these little animals. They come as an army, steadily and slowly advancing, always in the same direction, and "regardless of all obstacles, swimming across streams and even lakes of several miles in breadth, and committing considerable devastation on their line of march by the quantity of food they consume. In their turn they are pursued and harassed by crowds of beasts and birds of prey, as bears, wolves, foxes, dogs, wild cats, stoats, weasels, eagles, hawks, and owls, and never spared by man; even the domestic animals not usually predaceous, as cattle, foals, and reindeer, are said to join in the destruction, stamping them to the ground with their feet and even eating their bodies. Numbers also die from disease apparently produced from overcrowding. None ever return by the course by which they came, and the onward march of the survivors never ceases until they reach the sea, into which they plunge, and swimming onward in the same direction as before perish in the waves." One of these great migrations lasts for from one to three years. But it always ends in the total destruction of the migrating army. But the migration may be of advantage to the lemmings which remain in the original breeding grounds, leaving them with enough food, so that, on the whole, the migration results in gain to the species.

But most animals can not migrate to their betterment. In that case the only alternatives are adaptation or destruction. Some individuals by the possession of slight advantageous variations of structure are able to meet the new demands and survive, the rest die. The survivors produce young similarly advantageously different from the general type, and the adaptation increases with successive generations.

250. **Adjustment to surroundings a result of natural selection.**—To such causes as these we must ascribe the nice adjustment of each species to its surroundings. If a species or a group of individuals can not adapt itself to its environment, it will be crowded out by others that can do so. The former will disappear entirely from the earth, or else will be limited to surroundings with which it comes into perfect adjustment. A partial adjustment must with time become a complete one, for the individuals not adapted will be exterminated in the struggle for life. In this regard very small variations may lead to great results. A side issue apparently of little consequence may determine the fate of a species. Any advantage, no matter how small, will turn the scale of life in favor of its possessor and his progeny. "Battle within battle," says a famous naturalist, "must be continually recurring, with varying success. Yet in the long run the forces are so nicely balanced that the face of Nature remains for a long time uniform, though assuredly the merest trifle would give the victory to one organic being over another."

251. **Artificial selection.**—It has been long known that the nature of a herd or race of animals can be materially altered by a conscious selection on the part of man of these individuals which are to become parents. To "weed out" a herd artificially is to improve its blood. To select for reproduction the swiftest horses, the best milk cows, the most intelligent dogs, is to raise the standard of the herd or race in each of these respects by the simple action of heredity. Artificial selection has been called the "magician's wand," by which the breeder can summon up whatever animal form he will. If the parentage is chosen to a definite end, the process of heredity will develop the form

desired by a force as unchanging as that by which a stream turns a mill.

From the wild animals about him man has developed tho domestic animals which he finds useful The dog which man trains to care for his sheep is developed by selection from the most tractable progeny of the wolf which once devoured his flocks. By the process of artificial selection those individuals that are not useful to man or pleasing to his fancy have been destroyed, and those which contribute to his pleasure or welfare have been preserved and allowed to reproduce their kind. The various fancy breeds of pigeons—the carriers, pouters, tumblers, ruff-necks, and fan-tails—are all the descendants of the wild dove of Europe (*Columba livia*). These breeds or races or varieties have been produced by artificial selection. So it is with the various breeds of cattle and of hogs and of horses and dogs.

In this artificial selection new variations are more rapidly produced than in Nature by means of intercrossing different races, and by a more rapid weeding out of unfavorable—that is, of undesirable—variations. The rapid production of variations and the careful preservation of the desirable ones and rigid destruction of undesirable ones are the means by which many races of domestic animals are produced. This is artificial selection.

252. **Dependence of species on species.**—There was introduced into California from Australia, on young orange trees, a few years ago, an insect pest called the cottony cushion scale (*Icerya purchasi*). This pest increased in numbers with extraordinary rapidity, and in four or five years threatened to destroy completely the great orange orchards of California. Artificial remedies were of little avail. Finally, an entomologist was sent to Australia to find out if this scale insect had not some special natural enemy in its native country. It was found that in Australia a certain species of lady-bird beetle attacked and fed on the cottony

cushion scales and kept them in check. Some of these lady-birds (*Vedalia cardinalis*) were brought to California and released in a scale-infested orchard. The lady-birds, having plenty of food, thrived and produced many young. Soon the lady-birds were in such numbers that numbers of them could be distributed to other orchards. In two or three years the *Vedalias* had become so numerous and widely distributed that the cottony cushion scales began to diminish perceptibly, and soon the pest was nearly wiped out. But with the disappearance of the scales came also a disappearance of the lady-birds, and it was then discovered that the *Vedalias* fed only on cottony cushion scales and could not live where the scales were not. So now, in order to have a stock of *Vedalias* on hand in California it is necessary to keep protected some colonies of the cottony cushion scale to serve as food. Of course, with the disappearance of the predaceous lady-birds the scale began to increase again in various parts of the State, but with the sending of *Vedalias* to these localities the scale was again crushed. How close is the interdependence of these two species!

Similar relations can be traced in every group of animals. When the salmon cease to run in the Sacramento River in California the otter which feeds on them takes, it is said, to robbing the poultry-yards; and the bear, which also feeds on fish, strikes out for other game, taking fruit or chickens or bee-hives, whatever he may find.

CHAPTER XXI

ADAPTATIONS

253. Origin of adaptations.—The strife for place in the crowd of animals makes it necessary for each one to adjust itself to the place it holds. As the individual becomes fitted to its condition, so must the species as a whole. The species is therefore made up of individuals that are fitted or may become fitted for the conditions of life. As the stress of existence becomes more severe, the individuals fit to continue the species are chosen more closely. This choice is the automatic work of the conditions of life, but it is none the less effective in its operations, and in the course of centuries it becomes unerring. When conditions change, the perfection of adaptation in a species may be the cause of its extinction. If the need of a special fitness can not be met immediately, the species will disappear. For example, the native sheep of England have developed a long wool fitted to protect them in a cool, damp climate. Such sheep transferred to Cuba died in a short time, leaving no descendants. The warm fleece, so useful in England, rendered them wholly unfit for survival in the tropics. It is one advantage of man, as compared with other forms of life, that so many of his adaptations are external to his structure, and can be cast aside when necessity arises.

254. Classification of adaptations.—The various forms of adaptations may be roughly divided into five classes, as follows: (*a*) food securing, (*b*) self-protection, (*c*) rivalry, (*d*) defense of young, (*e*) surroundings.

The few examples which are given under each class,

some of them striking, some not especially so, are mostly
chosen from the vertebrates and from the insects, because
these two groups of animals are the groups with which beginning students of zoölogy are likely to be familiar, and
the adaptations referred to are therefore most likely to be
best appreciated. Quite as good and obvious examples could
be selected from any other groups of animals. The student

Fig. 156.—The deep-sea angler (*Corynolophus reinhardti*), which has a dorsal spine
modified to be a luminous "fishing rod and lure," attracting lantern-fishes
(*Echiostoma* and *Æthoprora*). An extraordinary adaptation for securing food.
(The angler is drawn after a figure of Lütken's.)

will find good practice in trying to discover examples shown
by the animals with which he may be familiar. That all
or any part of the body structure of any animal can be
called with truth an example of adaptation is plain from
what we know of how the various organs of the animal
body have come to exist. But by giving special attention
to such adaptations as are plainly obvious, beginning stu-

dents may be put in the way of independent observation along an extremely interesting and attractive line of zoölogical study.

255. **Adaptations for securing food.** — For the purpose of capture of their prey, some carnivorous animals are provided with strong claws, sharp teeth, hooked beaks, and other structures familiar to us in the lion, tiger, dog, cat, owl, and eagle. Insect-eating mammals have contrivances especially

FIG. 157.—The brown pelican, showing gular sac, which it uses in catching and holding fishes that form its food.

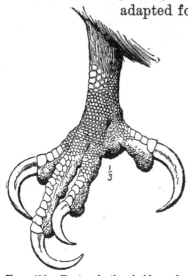

FIG. 158.—Foot of the bald eagle, showing claws for seizing its prey. (CHAPMAN.)

adapted for the catching of insects. The ant-eater, for example, has a curious, long sticky tongue which it thrusts forth from its cylindrical snout deep into the recesses of the ant-hill, bringing it out with its sticky surface covered with ants. Animals which feed on nuts are fitted with strong teeth or beaks for cracking them. Similar teeth are found in those fishes which feed on crabs, snails, or sea-urchins. Those mammals like the horse and cow, that feed on plants, have usually broad chisel-like incisor teeth for cutting off the foliage, and teeth of very similar form are developed in the different groups of plant-eating fishes. Molar teeth are found when it

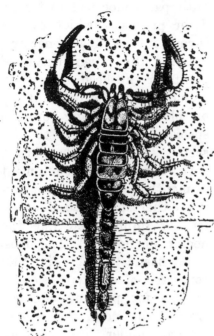

FIG. 159.—Scorpion, showing the special development of certain mouth parts (the maxillary palpi) as pincer-like organs for grasping prey. At the posterior tip of the body is the poisonous sting.

FIG. 160.—Head of mosquito (female), showing the piercing needle-like mouth parts which compose the "bill."

is necessary that the food should be crushed or chewed, and the sharp canine teeth go with a flesh diet. The long neck of the giraffe enables it to browse on the foliage of trees in grassless regions.

Insects like the leaf-beetles and the grasshoppers, that feed on the foliage of plants, have a pair of jaws, broad but

FIG. 161.—The praying-horse (*Mantis*) with fore legs developed as grasping organs.

sharply edged, for cutting off bits of leaves and stems. Those which take only liquid food, as the butterflies and sucking-bugs, have their mouth parts modified to form a slender, hollow sucking beak or proboscis, which can be

thrust into a flower nectary, or into the green tissue of plants or the flesh of animals, to suck up nectar or plant sap or blood, depending on the special food habits of the insect. The honey-bee has a very complicated equipment of mouth parts fitted for taking either solid food like pollen, or liquid food like the nectar of flowers. The mosquito has a "bill" (Fig. 160) composed of six sharp, slender needles for piercing and lacerating the flesh, and a long tubular under lip through which the blood can flow into the mouth. Some predaceous insects, as the praying-horse (Fig. 161), have their fore legs developed into formidable grasping organs for seizing and holding their prey.

256. **Adaptation for self-defense.**—For self-protection, carnivorous animals use the same weapons to defend themselves which serve to secure their prey; but these as well as

Fig. 162.—Acorns put into bark of tree by the Californian woodpecker (*Melanerpes formicivorus bairdii*).—From photograph, Stanford University, California.

other animals may protect themselves in other fashions. Most of the hoofed animals are provided with horns, struc-

FIG. 163.—Section of bark of live oak tree with acorns placed in it by the Californian woodpecker (*Melanerpes formicivorus bairdii*).—From photograph, Stanford University, California.

tures useless in procuring food but often of great effectiveness as weapons of defense. To the category of structures useful for self-defense belong the many peculiarities of coloration known as "recognition marks." These are marks, not otherwise useful, which are supposed to enable members of any one species to recognize their own kind among the mass of animal life. To this category belongs the

black tip of the weasel's tail, which remains the same whatever the changes in the outer fur. Another example is seen in the white outer feathers of the tail of the meadow-lark as well as in certain sparrows and warblers. The white on the skunk's back and tail serves the same purpose and also as a warning. It is to the skunk's advantage not to be hidden, for to be seen in the crowd of animals is to be avoided by them. The songs of birds and the calls of various creatures serve also as recognition marks. Each species knows and heeds its own characteristic song or cry, and it is a source of mutual protection. The fur-seal pup knows its mother's call, even though ten thousand other mothers are calling on the rookery.

FIG. 164.—Centiped. The foremost pair of legs is modified to be a pair of seizing and stinging organs. An adaptation for self-defense and for securing food.

The ways in which animals make themselves disagreeable or dangerous to their captors are almost as varied as the animals themselves. Besides the teeth, claws, and horns of ordinary attack and defense, we find among the mammals many special structures or contrivances which serve for defense through making their possession unpleasant. The scent glands of the skunk and its relatives are noticed above. The porcupine has the bristles in its fur specialized as quills, barbed and detachable. These quills fill the mouth of an attacking fox or wolf, and serve well the purpose of defense. The hedgehog of Europe, an animal of different nature, being related rather to the mole than to the squirrel, has a similar armature of quills. The armadillo of the tropics has movable shields, and when it withdraws its

FIG. 165.—Flying fishes. (The upper one a species of *Cypselurus*, the lower of *Exocœtus*.) These fishes escape from their enemies by leaping into the air and sailing or "flying" long distances.

head (which is also defended by a bony shield) it is as well protected as a turtle.

FIG. 166.—The horned toad (*Phrynosoma blainvillei*). The spiny covering repels many enemies.

Special organs for defense of this nature are rare among birds, but numerous among reptiles. The turtles are all

protected by bony shields, and some of them, the box-turtles, may close their shields almost hermetically. The
snakes broaden their heads, swell their necks, or show their
forked tongues to frighten their enemies. Some of them

Fig. 167.—Noki or poisonous scorpion-fish (*Emmydrichthys vulcanus*) with poisonous spines, from Tahiti.

are further armed with fangs connected with a venom gland,
so that to most animals their bite is deadly. Besides its
fangs the rattlesnake has a rattle on the tail made up of a

Fig. 168.—Mad tom (*Schilbeodes furiosus*) with poisoned pectoral spine.

succession of bony clappers, modified vertebræ, and scales,
by which intruders are warned of their presence. This
sharp and insistent buzz is a warning to animals of other
species and a recognition signal to those of its own kind.

Even the fishes have many modes of self-defense through giving pain or injury to those who would swallow them. The cat-fishes or horned pouts when attacked set immovably the sharp spine of the pectoral fin, inflicting a jagged wound. Pelicans who have swallowed a cat-fish have been known to die of the wounds inflicted by the fish's spine. In the group of scorpion-fishes and toad-fishes are certain genera in which these spines are provided with poison glands. These may inflict very severe wounds to other fishes, or even to birds or man. One of this group of poison-fishes is the noki (*Emmydrich-thys*, Fig. 167). A group of small fresh-water cat-fishes, known as the mad toms (Fig. 168), have also a poison gland attached to the pectoral spine, and its sting is most exasperating, like the sting of a wasp. The sting-rays (Fig. 169) of many species have a strong, jagged spine on the tail, covered with slime, and armed with broad saw-like teeth. This inflicts a dangerous wound, not through the presence of specific venom, but from the danger of blood poisoning arising from the slime, and the ragged or unclean cut.

Fig. 169.—A sting-ray (*Urolophus goodei*), from Panama.

Many fishes are defended by a coat of mail or a coat of sharp thorns. The globe-fishes and porcupine-fishes (Fig. 170) are for the most part defended by spines, but their instinct to swallow air gives them an additional safeguard. When one of these fishes is disturbed it rises to the surface, gulps air until its capacious stomach is filled, and then floats belly upward on the surface. It is thus protected from other fishes, though easily taken by man. The torpedo, olcctric eel, eleclric cat-fish, and star-gazer, surprise and stagger their captors by means of electric shocks. In the torpedo or electric ray (Fig. 171), found on the sandy shores of all warm seas, on either side of the head is a large honeycomb-like structure which yields a strong electric shock whenever the live fish is touched. This shock is felt severely if the fish be stabbed with a knife or metallic spear. The electric eel of the rivers of Paraguay and southern Brazil is said to give severe shocks to herds of wild horses driven through the streams, and similar accounts are given of the electric cat-fish of the Nile.

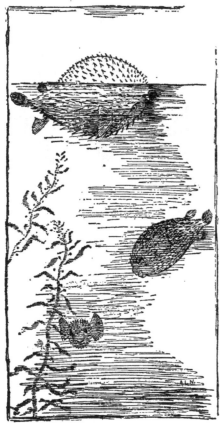

FIG. 170.—Porcupine-fish (*Diodon hystrix*), the lower ones swimming normally, the upper one floating belly upward, with inflated stomach.—Drawn from specimens from the Florida Keys.

Among the insects, the possession of stings is not uncommon. The wasps and bees are familiar examples of stinging insects, but many other kinds, less familiar, are similarly protected. All insects have their bodies covered with a coat of armor, composed of a horny substance called chitin. In some cases this chitinous coat is very thick and serves to protect them effectually. This is especially true of the beetles. Some insects are inedible (as mentioned in Chapter XXIV), and are conspicuously colored so as to be readily recognized by insectivorous birds. The birds, knowing by experience that these insects are ill-tasting, avoid them. Others are effectively concealed from their enemies by their close resemblance in color and marking to their surroundings. These protective resemblances are discussed in Chapter XXIV.

257. **Adaptation for rivalry.**—In questions of attack and defense, the need of meeting animals of their own kind as well as animals of other races must be considered. In struggles of species with those of their own kind, the term rivalry may be applied. Actual warfare is confined mainly to males in the breeding season and to polygamous animals. Among those in which the male mates with many females, he must struggle with other males for their possession. In all the groups of vertebrates the sexes are about equal in numbers. Where mating exists, either for the season or for life, this condition does not

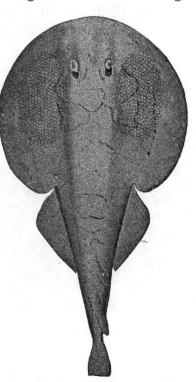

Fig. 171.—Torpedo or electric ray (*Narcine brasiliensis*), showing electric cells.

involve serious struggle or destructive rivalry.

Among monogamous birds, or those which pair, the male courts the female of his choice by song and by display of his bright feathers. The female consents to be chosen by the one which pleases her. It is believed that the handsomest, most vivacious, and most musical males are the ones most successful in such courtship. With polygamous animals there is intense rivalry among the males in the mating season, which in almost all species is in the spring. The strongest males survive and reproduce their strength. The most notable adaptation is seen in the superior size of teeth, horns, mane, or spurs. Among the polygamous fur seals and sea lions the male is about four times

Fig. 172.—A wild duck (*Aythya*) family. Male, female, and præcocial young.

the size of the female. In the polygamous family of deer, buffalo, and the domestic cattle and sheep, the male is larger and more powerfully armed than the female. In the polygamous group to which the hen, turkey, and peacock belong the males possess the display of plumage, and the structures adapted for fighting, with the will to use them.

258. **Adaptations for the defense of the young.**—The protection of the young is the source of many adaptive structures as well as of the instincts by which such structures are

utilized.   In general, those animals are highest in development, with best means of holding their own in the struggle

FIG. 173.—The altricial nestlings of the Blue jay (*Cyanocitta cristata*).

for life, that take best care of their young.   The homes
of animals are specially discussed in the volume on Ani-

mal Life, but those instincts which lead to home-building
may all be regarded as useful adaptations in preserving
the young.   Among the lower or more coarsely organized

FIG. 174.—Kangaroo (*Macropus rufus*) with young in pouch.

birds, such as the chicken, the duck, and the auk, as with the reptiles, the young animal is hatched with well-developed muscular system and sense organs, and is capable of running about, and, to some extent, of feeding itself. Birds of this type are known as *præcocial* (Fig. 172), while the name *altricial* (Fig. 173) is applied to the more highly organized forms, such as the thrushes, doves, and song-birds generally. With these the young are hatched in a wholly helpless condition, with ineffective muscles, deficient senses, and dependent wholly upon the parent. The altricial condition demands the building of a nest, the establishment of a home, and the continued care of one or both of the parents.

Fig. 175.—Egg-case of California barn-door skate (*Raja binoculata*) cut open to show young inside. (Young issues naturally at one end of the case.)

The very lowest mammals known, the duck-bills (Monotremes) of Australia, lay large eggs in a strong shell like those of a turtle, and guard them with great jealousy. But with almost all mammals the egg is very small and without much food-yolk. The egg begins its development within the body. It is nourished by the blood of the mother, and after birth the young is cherished by her, and fed by milk secreted by specialized glands of the skin. All these features are adaptations tending toward the preservation of the young. In the

Fig. 176.—Egg-case of the cockroach.

division of mammals next lowest to the Monotremes—the kangaroo, opossum, etc.—the young are born in a very immature state and are at once seized by the mother and thrust into a pouch or fold of skin along the abdomen, where they are kept until they are able to take care of themselves (Fig. 174). This is an interesting and ingenious

adaptation, but less specialized and less perfect an adaptation than the conditions found in ordinary mammals.

Among the insects, the special provisions for the protection and care of the eggs and the young are wide-spread and various. Some of those adaptations which take the special form of nests or "homes" are described in the volume on Animal Life. The eggs of the common cockroach are laid in small packets inclosed in a firm wall (Fig.

FIG. 177. — Giant water-bug (*Serphus*). Male carrying eggs on its back.

176). The eggs of the great water-bugs are carried on the back of the male (Fig. 177); and the spiders lay their eggs in a silken sac or cocoon, and some of the ground or

FIG. 178.—Cocoon inclosing the pupa of the great Ceanothus moth. Spun of silk by the larva before pupation.

running spiders (*Lycosidæ*) drag this egg-sac, attached to the tip of the abdomen, about with them. The young spiders when hatched live for some days inside this sac, feeding on each other! Many insects have long, sharp piercing ovipositors, by means of which the eggs are deposited in the ground or in the leaves or stems of green plants, or even in the hard wood of tree-trunks. Some of

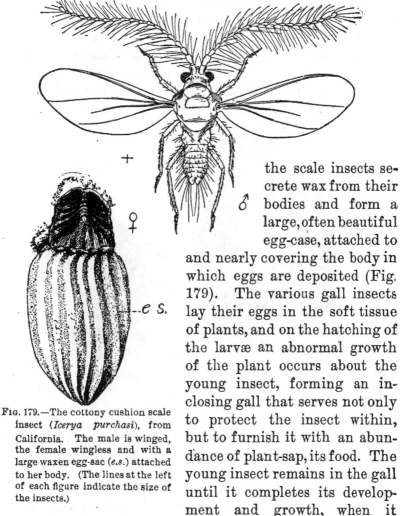

FIG. 179.—The cottony cushion scale insect (*Icerya purchasi*), from California. The male is winged, the female wingless and with a large waxen egg-sac (*e.s.*) attached to her body. (The lines at the left of each figure indicate the size of the insects.)

the scale insects secrete wax from their bodies and form a large, often beautiful egg-case, attached to and nearly covering the body in which eggs are deposited (Fig. 179). The various gall insects lay their eggs in the soft tissue of plants, and on the hatching of the larvæ an abnormal growth of the plant occurs about the young insect, forming an inclosing gall that serves not only to protect the insect within, but to furnish it with an abundance of plant-sap, its food. The young insect remains in the gall until it completes its development and growth, when it gnaws its way out. Such insect galls are especially abundant on oak trees (Fig. 180). The care of the eggs and the young of the social insects, as the bees and ants, are described in Chapter XXII.

259. Adaptations concerned with surroundings in life.—A large part of the life of the animal is a struggle with the environment itself; in this struggle only those that are adapted live and leave descendants fitted like themselves. The fur of mammals fits them to their surroundings. As the fur differs, so may the habits change. Some animals are active in winter; others, as the bear, hibernate, sleeping in caves or hollow trees or in burrows until conditions are favorable for their activity. Most snakes and lizards hibernate in cold weather. In the swamps of Louisiana,

Fig. 180.—The giant gall of the white oak (California), made by the gall insect *Andri- eus californicus*. The gall at the right cut open to show tunnels made by the insects in escaping from the gall.—From photograph.

in winter, the bottom may often be seen covered with water snakes lying as inert as dead twigs. Usually, however, hibernation is accompanied by concealment. Some animals in hibernation may be frozen alive without apparent injury. The blackfish of the Alaska swamps, fed to dogs when frozen solid, has been known to revive in the heat of the dog's stomach and to wriggle out and escape. As animals resist heat and cold by adaptations of structure or habits, so may they resist dryness. Certain fishes hold reservoirs of water above their gills, by means of which they can breathe during short excursions from the water. Still others (mud-fishes) retain the primitive lung-like structure of the swim-bladder, and are able to breathe air when, in the dry season, the water of the pools is reduced to mud.

Another series of adaptations is concerned with the places chosen by animals for their homes. The fishes that

live in water have special organs for breathing under water (Fig. 182). Many of the South American monkeys have the tip of the tail adapted for clinging to limbs of trees or to the bodies of other monkeys of its own kind. The hooked claws of the bat hold on to rocks, the bricks of chimneys, or to the surface of hollow trees where the bat sleeps through the day. The tree-frogs (Fig. 183) or tree-toads have the tips of the toes swollen, forming little pads by which they cling to the bark of trees.

Among other adaptations relating to special surroundings or conditions of life are the great cheek pouches of the pocket gophers, which carry off the soil dug up by the large shovel-like feet when the gopher excavates its burrow.

Fig. 181.—Insect galls on leaf.

Those insects which live underground, making burrows or tunnels in the soil, have their legs or other parts adapted for digging and burrowing. The mole cricket (Fig. 184) has its legs stout and short, with broad, shovel-like feet. Some water-beetles (Fig. 185) and water-bugs have one or more of the pairs of legs flattened and broad to serve as oars or paddles for swimming. The grasshoppers or locusts, who leap,

have their hind legs greatly enlarged and elongated, and provided with strong muscles, so as
to make of them "leaping legs." The grubs

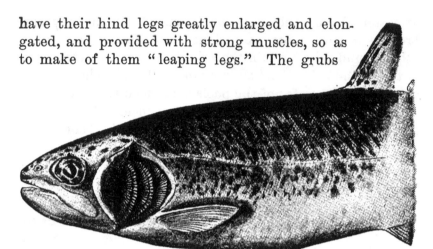

FIG. 182.—Head of rainbow trout (*Salmo irideus*) with gill cover bent back to show
gills, the breathing organs.

or larvæ of beetles which live as "borers" in tree-trunks
have mere rudiments of legs, or none at all (Fig. 186).
They have great, strong, biting jaws for cutting away
the hard wood. They move simply by wriggling along
in their burrows or tunnels.

Insects that live
in water either come
up to the surface to
breathe or take down
air underneath their
wings, or in some
other way, or have
gills for breathing the
air which is mixed
with the water. These
gills are special adap-

FIG. 183.—Tree-toad (*Hyla regilla*).

tive structures which present a great variety of form and
appearance. In the young of the May-flies they are delicate plate-like flaps projecting from the sides of the body.
They are kept in constant motion, gently waving back and

forth in the water so as to maintain currents to bring fresh water in contact with them. Young mosquitoes (Fig. 187) do not have gills, but come up to the surface to breathe. The larvæ, or wrigglers, breathe through a special

Fig. 184.—The mole cricket (*Gryllotalpa*), with fore feet modified for digging.

Fig. 185.—A water-beetle (*Hydrophilus*).

tube at the posterior tip of the body, while the pupæ have a pair of horn-like tubes on the back of the head end of the body.

260. **Degree of structural change in adaptations.**—While among the higher or vertebrate animals, especially the fishes and reptiles, most remarkable cases of adaptations occur, yet the structural changes are for the most part external, never seriously affecting the development of the internal organs other than the skeleton. The organization of these higher animals is much less plastic than among the invertebrates. In

Fig. 186.—Wood-boring beetle larva (*Prionus*).

general, the higher the type the more persistent and **un**changeable are those structures not immediately exposed to the influence of the struggle for existence. It is thus the outside of an animal that tells where its ancestors have lived. The inside, suffering little change, whatever the surroundings, tells the real nature of the animal.

261. **Vestigial organs.**—In general, all the peculiarities of animal structure find their explanation in some need of adaptation. When this need ceases, the structure itself tends to disappear or else to serve some other need. In the bodies of most animals there are certain incomplete or rudimentary organs or structures which serve no distinct useful purpose. They are structures which, in the ancestors of the animals now possessing them, were fully developed functional organs, but which, because of a change in habits or conditions of living, are of no further need, and are gradually dying out. Such organs are called vestigial organs. Examples are the disused ear muscles of man, the vermiform appendix in

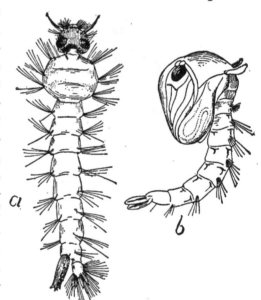

Fig. 187.—Young stages of the mosquito. *a*, larva (wriggler) ; *b*, pupa.

man, which is the reduced and now useless anterior end of the large intestine. In the lower animals, the thumb or degenerate first finger of the bird with its two or three little quills serves as an example. So also the reduced and elevated hind toe of certain birds, the splint bones or rudimentary side toes of the horse, the rudimentary eyes of blind fishes, the minute barbel or beard of the horned dace or chub, and the rudimentary teeth of the right whales and sword-fish.

Each of these vestigial organs tells a story of some past adaptation to conditions, one that is no longer needed in the life of the species. They have the same place in the study of animals that silent letters have in the study of words. For example, in our word knight the *k* and *gh* are no longer sounded; but our ancestors used them both, as the Germans do to-day in their cognate word *Knecht*. So with the French word *temps*, which means time, in which both *p* and *s* are silent. The Romans, from whom the French took this word, needed all its letters, for they spelled and pronounced it *tempus*. In general, every silent letter in every word was once sounded. In like manner, every vestigial structure was once in use and helpful or necessary to the life of the animal which possessed it.

Horns of two male elk interlocked while fighting.
(Permission of G. O. SHIELDS, publisher of Recreation.)

CHAPTER XXII

ANIMAL COMMUNITIES AND SOCIAL LIFE

262. Man not the only social animal.—Man is commonly called the social animal, but he is not the only one to which this term may be applied. There are many others which possess a social or communal life. A moment's thought brings to mind the familiar facts of the communal life of the honey-bee and of the ants. And there are many other kinds of animals, not so well known to us, that live in communities or colonies, and live a life which in greater or less degree is communal or social. In this connection we may use the term communal for the life of those animals in which the division of labor is such that the individual is dependent for its continual existence on the community as a whole. The term social life would refer to a lower degree of mutual aid and mutual dependence.

263. The honey-bee.—Honey-bees live together, as we know, in large communities. We are accustomed to think of honey-bees as the inhabitants of bee-hives, but there were bees before there were hives. The "bee-tree" is familiar to many of us. The bees, in Nature, make their home in the hollow of some dead or decaying tree-trunk, and carry on there all the industries which characterize the busy communities in the hives. A honey-bee community comprises three kinds of individuals (Fig. 188)—namely, a fertile female or queen, numerous males or drones, and many infertile females or workers. These three kinds of individuals differ in external appearance sufficiently to be readily recognizable. The workers are

314

smaller than the queens and drones, and the last two differ
in the shape of the abdomen, or hind body, the abdomen of
the queen being longer and more slender than that of the

Fig. 188.—Honey-bee. *a*, drone or male ; *b*, worker or infertile female ; *c*, queen or
fertile female.

male or drone.   In a single community there is one queen,
a few hundred drones, and ten to thirty thousand workers.
The number of drones and workers varies at different
times of the year, being smallest in winter.   Each kind of
individual has certain work or business to do for the whole
community.   The queen lays all the eggs from which new
bees are born; that is, she is the mother of the entire
community.   The drones or males have simply to act as
royal consorts; upon them depends the fertilization of the
eggs.   The workers undertake all the food-getting, the
care of the young bees, the comb-building, the honey-mak-
ing—all the industries with which we are more ·or less
familiar that are carried on in the hive.   And all the
work done by the workers is strictly work for the whole
community; in no case does the worker bee work for itself
alone; it works for itself only in so far as it is a member
of the community.

How varied and elaborately perfected these industries
are may be perceived from a brief account of the life history of a bee community.   The interior of the hollow in
the bee-tree or of the hive is filled with " comb "—that is,
with wax molded into hexagonal cells and supports for
these cells.   The molding of these thousands of symmet-

rical cells is accomplished by the workers by means of their specially modified trowel-like mandibles or jaws. The wax itself, of which the cells are made, comes from the bodies of the workers in the form of small liquid drops which exude from the skin on the under side of the abdomen or hinder body rings. These droplets run together, harden and become flattened, and are removed from the wax plates, as the peculiarly modified parts of the skin which produce the wax are called, by means of the hind legs, which are furnished with scissor-like contrivances for cutting off the wax (Fig. 189). In certain of the cells are stored the pollen and honey, which serve as food for the community. The pollen is gathered by the workers from certain favorite flowers and is carried by them from the flowers to the hive in the "pollen baskets," the slightly concave outer surfaces of one of the segments of the broadened and flattened hind legs. This concave surface is lined on each margin with a row of incurved stiff hairs which hold the pollen mass securely in place (Fig. 189). The "honey" is the nectar of flowers which has been sucked up by the workers by means of their elaborate lapping and sucking mouth parts and swallowed into a sort of honey-sac or stomach, then brought to the hive and regurgitated into the

Fig. 189.—Posterior leg of worker honey-bee. The concave surface of the upper large joint with the marginal hairs is the pollen basket; the wax shears are the cutting surfaces of the angle between the two large segments of the leg.

cells. This nectar is at first too watery to be good honey, so the bees have to evaporate some of this water. Many of the workers gather above the cells containing nectar, and buzz—that is, vibrate their wings violently. This creates currents of air which pass over the exposed nectar and increase the evaporation of the water. The violent buzzing raises the temperature of the bees' bodies, and this warmth given off to the air also helps make evaporation more rapid. In addition to bringing in food the workers also bring in, when necessary, "propolis," or the resinous gum of certain trees, which they use in repairing the hive, as closing up cracks and crevices in it.

In many of the cells there will be found, not pollen or honey, but the eggs or the young bees in larval or pupal condition (Fig. 190).

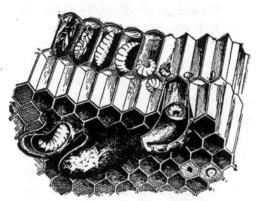

FIG. 190.—Cells containing eggs, larvæ, and pupæ of the honey-bee. The lower large, irregular cells are queen cells.—After BENTON.

The queen moves about through the hive, laying eggs. She deposits only one egg in a cell. In three days the egg hatches, and the young bee appears as a helpless, soft, white, footless grub or larva. It is cared for by certain of the workers, that may be called nurses. These nurses do not differ structurally from the other workers, but they have the special duty of caring for the helpless young bees. They do not go out for pollen or honey, but stay in the hive. They are usually the new bees—i. e., the youngest or most recently added workers. After they act as nurses for a week or so they take their places with the food-gathering workers, and other new bees act as nurses. The nurses feed the young or larval bees at first with a highly nutritious food called bee-jelly, which the nurses make in their stomach, and regurgitate for the larvæ. After the larvæ are two or three days old they are fed with pollen and honey. Finally, a small mass of food is put into the cell, and the cell is "capped" or covered with wax. The larva, after eating all the food, in two or three days more changes into a pupa, which lies quiescent without eating for thirteen days, when it changes into a full-grown bee. The new bee breaks open the cap of the cell with its jaws, and comes out into the hive, ready to take up its share of the work for the community. In a few cases, however, the life history is different. The nurses will tear down several cells around some single one, and enlarge this inner one into a great irregular vase-shaped cell. When the egg hatches, the grub or larva is fed bee-jelly as long as it remains a larva, never being given ordinary pollen and honey at all. This larva finally pupates, and there issues from the pupa not a worker or drone bee, but a new queen. The egg from which the queen is produced is the same as the other eggs, but the worker nurses by feeding the larva only the highly nutritious bee-jelly make it certain that the new bee shall become a queen instead of a worker. It is also to be noted that the male bees or drones are hatched from eggs that are not fertilized, the queen having it in her power to lay either fertilized or unfertilized eggs. From the fertilized eggs hatch larvæ which develop into queens or workers, depending on the manner of their nourishment; from the unfertilized eggs hatch the males.

When several queens appear there is much excitement in the community. Each community has normally a single one, so that when additional queens appear some rearrangement is necessary. This rearrangement comes about first by fighting among the queens until only one of the new queens is left alive. Then the old or mother queen issues from the hive or tree followed by many of the workers. She and her followers fly away together, finally alighting on some tree branch and massing there in a dense swarm. This is the familiar phenomenon of "swarming." The swarm finally finds a new hollow tree, or in the case of the hive-bee (Fig. 191) the swarm is put into a new hive, where the bees build cells, gather food, produce young, and thus

Fig. 191.—Hiving a swarm of honey-bees. Photograph by S. J. HUNTER.

found a new community. This swarming is simply an emigration, which results in the wider distribution and in the increase of the number of the species. It is a peculiar but effective mode of distributing and perpetuating the species.

There are many other interesting and suggestive things which might be told of the life in a bee community: how the community protects itself from the dangers of starvation when food is scarce or winter comes on by killing the useless drones and the immature bees in egg and larval stage; how the instinct of home-finding has been so highly developed that the worker bees go miles away for honey and nectar, flying with unerring accuracy back to the hive; of the extraordinarily nice structural modifications which adapt the bee so perfectly for its complex and varied businesses; and of the tireless persistence of the workers until they fall exhausted and dying in the performance of their duties. The community, it is important to note, is a persistent or continuous one. The workers do not live long, the spring broods usually not over two or three months, and the fall broods not more than six or eight months; but new ones are hatching while the old ones are dying, and the community as a whole always persists. The queen may live several years, perhaps as many as five.* She lays about one million eggs a year.

264. **The ants.**—There are many species of ants, two thousand or more, and all of them live in communities and show a truly communal life. There is much variety of habit in the lives of different kinds of ants, and the degree in which the communal or social life is specialized or elaborated varies much. But certain general conditions prevail in the life of all the different kinds of individuals—

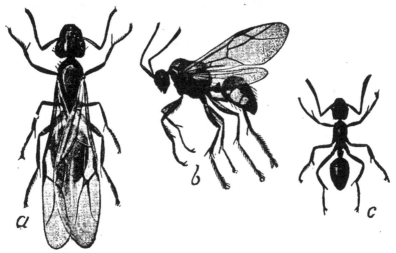

Fig. 192.—Female (*a*), male (*b*), and worker (*c*) of an ant (*Camponotus* sp.).

sexually developed males and females that possess wings, and sexually undeveloped workers that are wingless (Fig. 192). In some kinds the workers show structural differ-

---

* A queen bee has been kept alive for fifteen years,

ences among themselves, being divided into small workers, large workers, and soldiers. The workers are not, as with the bees, all infertile females, but they are both male and female, both being infertile. Although the life of the ant communities is much less familiar and fully known than that of the bees, it is even more remarkable in its specializations and elaborateness. The ant home, or nest, or formicary, is, with most species, a very elaborate underground, many-storied labyrinth of galleries and chambers. Certain rooms are used for the storage of food; certain others as "nurseries" for the reception and care of the young; and others as stables for the ants' cattle, certain plant-lice or scale-insects which are sometimes collected and cared for by the ants. The food of ants comprises many kinds of vegetable and animal substances, but the favorite food, or "national dish," as it has been called, is a sweet fluid which is produced by certain small insects, the plant-lice (Aphidæ) and scale-insects (Coccidæ). These insects live on the sap of plants; rose-bushes are especially favored with their presence. The worker ants (and we rarely see any ants but the wingless workers, the winged males and females appearing out of the nest only at mating time) find these honey-secreting insects, and gently touch or stroke them with their feelers (antennæ), when the plant-lice allow tiny drops of the honey to issue from the body, which are eagerly drunk by the ants. It is manifestly to the advantage of the ants that the plant-lice should thrive; but they are soft-bodied, defenseless insects, and readily fall a prey to the wandering predaceous insects like the lady-birds and aphis lions. So the ants often guard small groups of plant-lice, attacking, and driving away the would-be ravagers. When the branch on which the plant-lice are gets withered and dry, the ants have been observed to carry the plant-lice carefully to a fresh, green branch. In the Mississippi Valley a certain kind of plant-louse lives on the roots of corn. Its eggs are deposited in the ground in the autumn and hatch the following spring before the corn is planted. Now, the common little brown ant lives abundantly in the corn-fields, and is specially fond of the honey secreted by the corn-root plant-louse. So, when the plant-lice hatch in the spring before there are corn roots for them to feed on, the little brown ants with great solicitude carefully place the plant-lice on the roots of a certain kind of knotweed which grows in the field, and protect them until the corn ger-'minates. Then the ants remove the plant-lice to the roots of the corn, their favorite food plant. In the arid lands of New Mexico and Arizona the ants rear their scale-insects on the roots of cactus. Other kinds of ants carry plant-lice into their nests and provide them with food there. Because the ants obtain food from the plant-lice and take care of them, the plant-lice are not inaptly called the ants' cattle.

Like the honey-bees, the young ants are helpless little grubs or larvæ, and are cared for and fed by nurses. The so-called ants' eggs, little white, oval masses, which we often see being carried in the mouths of ants in and out of an ants' nest, are not eggs, but are the pupæ which are being brought out to enjoy the warmth and light of the sun or being taken back into the nest afterward.

In addition to the workers that build the nest and collect food and care for the plant-lice, there is in many species of ants a kind of individuals called soldiers. These are wingless, like the workers, and are also, like the workers, not capable of laying or of fertilizing eggs. It is the business of the soldiers, as their name suggests, to fight. They protect the community by attacking and driving away predaceous insects, especially other ants. The ants are among the most warlike of insects. The soldiers of a community of one species of ant often sally forth and attack a community of some other species. If successful in battle the workers of the victorious community take possession of the food stores of the conquered and carry

them to their own nest.   Indeed, they go even further; they
may make slaves of the conquered ants.   There are numerous species of the so-called slave-making ants.   The slave-
makers carry into their own nest the eggs and larvæ and
pupæ of the conquered community, and when these come
to maturity they act as slaves of the victors—that is, they
collect food, build additions to the nests, and care for the
young of the slave-makers.   This specialization goes so far
in the case of some kinds of ants, like the robber-ant of
South America (*Eciton*), that all of the *Eciton* workers have
become soldiers, which no longer do any work for themselves.   The whole community lives, therefore, wholly by
pillage or by making slaves of other kinds of ants.   There
are four kinds of individuals in a robber-ant community—
winged males, winged females, and small and large wingless soldiers.   There are many more of the small soldiers
than of the large, and some naturalists believe that the few
latter, which are distinguished by heads and jaws of great
size, act as officers.   On the march the small soldiers are
arranged in a long, narrow column, while the large soldiers
are scattered along on either side of the column and appear
to act as sentinels and directors of the army.   The observations made by the famous Swiss students of ants, Huber
and Forel, and by other naturalists, read like fairy tales,
and yet are the well-attested and often reobserved actual
phenomena of the extremely specialized communal and
social life of these animals.

265. **Other communal insects.**—The termites or white
ants (not true ants) are communal insects.   Some species
of termites in Africa live in great mounds of earth, often
fifteen feet high.   The community comprises hundreds of
thousands of individuals, which are of eight kinds (Fig. 193),
viz., sexually active winged males, sexually active winged
females, other fertile males and females which are wingless,
wingless workers of both sexes not capable of reproduction, and wingless soldiers of both sexes also incapable of

reproduction. The production of new individuals is the sole business of the fertile males and females; the workers build the nest and collect food, and the soldiers protect the community from the attacks of marauding insects. The egg-laying queen grows to monstrous size, being sometimes

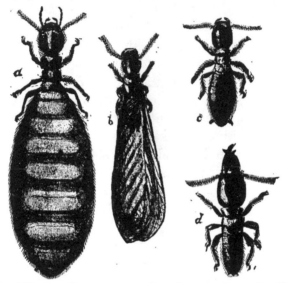

FIG. 193.—Termites. *a*, queen; *b*, male; *c*, worker; *d*, soldier.

five or six inches long, while the other individuals of the community are not more than half or three quarters of an inch long. The great size of the queen is due to the enormous number of eggs in her body.

The bumble-bees live in communities, but their social arrangements are very simple ones compared with those of the honey-bee. There is, in fact, among the bees a series of gradations from solitary to communal life. The interesting little green carpenter-bees live a truly solitary life. Each female bores out the pith from five or six inches of an elder branch or raspberry cane, and divides this space into a few cells by means of transverse partitions (Fig. 194). In each cell she lays an egg, and puts with it enough food —flower pollen—to last the grub or larva through its life.

She then waits in an upper cell of the nest until the young bees issue from their cells, when she leads them off, and each begins active life on its own account.   The mining-

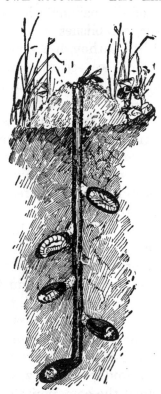

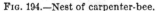

FIG. 194.—Nest of carpenter-bee.   FIG. 195.—Nest of *Andrena*, the mining-bee.

bees (*Andrena*), which make little burrows (Fig. 195) in a clay bank, live in large colonies—that is, they make their nest burrows close together in the same clay bank, but each female makes her own burrow, lays her own eggs in it, furnishes it with food—a kind of paste of nectar and pollen—and takes no further care of her young.   Nor has she at any time any special interest in her neighbors.   But with the smaller mining-bees, belonging to the genus *Halictus*, several females unite in making a common burrow, after which each female makes side passages of her own, extending from the main or public entrance burrow. As a well-known entomologist has said, *Andrena* builds villages composed of individual homes, while *Halictus* makes cities composed of apartment houses. The bumble-bee (Fig. 196), however, establishes a real community with a truly communal life, although a very simple one. The few bumble-bees which we see in winter time are queens; all other bumble-bees die in the autumn. In the spring a queen selects some deserted nest of a field-mouse, or a hole in the ground, gathers pollen which she molds into a rather large irregular mass and puts into the hole, and lays a few eggs on the pollen mass. The young grubs or larvæ which soon hatch feed on the pollen, grow, pupate, and issue as workers—winged bees a little smaller than the queen. These workers bring more pollen, enlarge the nest, and make irregular cells in the pollen mass, in each of which the queen lays an egg. She gathers no more pollen, does no more work except that of egg-laying. From these new eggs are produced more workers, and so on until the community may come to be pretty large. Later in the summer males and females are produced and mate. With the approach of winter all the workers and males die, leaving only the fertilized females, the queens, to live through the winter and found new communities in the spring.

Fig. 196.—Bumble-bees. *a*, worker; *b*, queen or fertile female.

The social wasps show a communal life like that of the bumble-bees. The only yellow-jackets and hornets that live through the winter are fertilized females or queens.

When spring comes each queen builds a small nest suspended from a tree branch, and consisting of a small comb inclosed in a covering or envelope open at the lower end. The nest is composed of "wasp paper," made by chewing bits of weather-beaten wood taken from old fences or outbuildings. In each of the cells the queen lays an egg. She deposits in the cell a small mass of food, consisting of some chewed insects or spiders. From these eggs hatch grubs which eat the food prepared for them, grow, pupate, and issue as worker bees, winged and slightly smaller than the queen (Fig. 197). The workers enlarge the nest, adding more combs and making many cells, in each of which the queen lays an egg. The workers provision the cell with chewed insects, and other broods of workers are

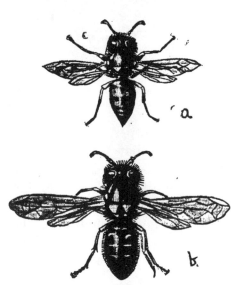

rapidly hatched. The community grows in numbers and the nest grows in size until it comes to be the great ball-like oval mass which we know so well as a hornets' nest (Figs. 198 and 199), a thing to be left untouched. Sometimes the nest is built underground. When disturbed, they swarm out of the hole and fiercely attack any invading foe in sight. After a number of broods of workers has

Fig. 197.—The yellow-jacket (*Vespa*), a social wasp. *a*, worker; *b*, queen.

been produced, broods of males and females appear and mating takes place. In the late fall the males and all of the many workers die, leaving only the new queens to live through the winter.

The bumble-bees and social wasps show an intermediate condition between the simply gregarious or neighborly

FIG. 198.—Nest of *Vespa*, a social wasp. From photograph.

FIG. 199.—Nest of *Vespa* opened to show combs within.

mining-bees and the highly developed, permanent honey-bee community. Naturalists believe that the highly organized communal life of the honey-bees and the ants is a development from some simple condition like that of the bumble-bees and social wasps, which in its turn has grown out of a still simpler, mere gregarious assembly of the individuals of one species. It is not difficult to see how such a development could in the course of a long time take place.

266. **Gregariousness and mutual aid.**—The simplest form of social life is shown among those kinds of animals in which many individuals of one species keep together, forming a great band or herd. In this case there is not much division of labor, and the safety of the individual is not wholly bound up in the fate of the herd. Such animals are

said to be *gregarious* in habit. The habit undoubtedly is advantageous in the mutual protection and aid afforded the individuals of the band. This mutual help in the case of many gregarious animals is of a very positive and obvious character. In other cases this gregariousness is reduced to a matter of slight or temporary convenience, possessing but little of the element of mutual aid. The great herds of reindeer in the north, and of the bison or buffalo which once ranged over the Western American plains, are examples of a gregariousness in which mutual protection from enemies, like wolves, seems to be the principal advantage gained. The bands of wolves which hunted the buffalo show the advantage of mutual help in aggression as well as in protection. In this banding together of wolves there is active co-operation among individuals to obtain a common food supply. What one wolf can not do—that is, tear down a buffalo from the edge of the herd—a dozen can do, and all are gainers by the operation. On the other hand, the vast assembling of sea-birds (Fig. 200) on certain ocean islands and rocks is a condition probably brought about rather by the special suitableness of a few places for safe breeding than from any special mutual aid afforded; still, these sea-birds undoubtedly combine to drive off attacking eagles and hawks. Eagles are usually considered to be strictly solitary in habit (the unit of solitariness being a pair, not an individual); but the description, by a Russian naturalist, of the hunting habits of the great white-tailed eagle (*Haliætos albicilla*) on the Russian steppes shows that this kind of eagle at least has adopted a gregarious habit, in which mutual help is plainly obvious. This naturalist once saw an eagle high in the air, circling slowly and widely in perfect silence. Suddenly the eagle screamed loudly. "Its cry was soon answered by another eagle, which approached it, and was followed by a third, a fourth, and so on, till nine or ten eagles came together and soon disappeared." The naturalist, following them, soon discovered them gathered

Fig. 200.—Pallas's murres (*Uria lomvia arra*) assembled on Walrus Island, one of the Pribilof group in Bering Sea. Photograph by HARRY CHICHESTER, photographer for the Fur Seal Commission.

about the dead body of a horse. The food found by the first was being shared by all. The association of pelicans in fishing is a good example of the advantage of a gregarious and mutually helpful habit. The pelicans sometimes go fishing in great bands, and, after having chosen an appropriate place near the shore, they form a wide half-circle facing the shore, and narrow it by paddling toward the land, catching the fish which they inclose in the ever-narrowing circle.

The wary Rocky Mountain sheep live together in small bands, posting sentinels whenever they are feeding or resting, who watch for and give warning of the approach of enemies. The beavers furnish a well-known and very interesting example of mutual help, and they exhibit a truly communal life, although a simple one. They live in " villages " or communities, all helping to build the dam across the stream, which is necessary to form the broad marsh or pool in which the nests or houses are built. Prairiedogs live in great villages or communities which spread over many acres. They tell each other by shrill cries of the approach of enemies, and they seem to visit each other and to enjoy each other's society a great deal, although that they afford each other much actual active help is not apparent. Birds in migration are gregarious, although at other times they may live comparatively alone. In their long flights they keep together, often with definite leaders who seem to discover and decide on the course of flight for the whole great flock. The wedge-shaped flocks of wild geese flying high and uttering their sharp, metallic call in their southward migrations are well known in many parts of the United States. Indeed, the more one studies the habits of animals the more examples of social life and mutual help will be found. Probably most animals are in some degree gregarious in habit, and in all cases of gregariousness there is probably some degree of mutual aid.

267. **Division of labor and communal life.**—It has been explained in Animal Life that the complexity of the bodies of the higher animals depends on a specialization or differentiation of parts, due to the assumption of different functions or duties by different parts of the body; that the degree of structural differentiation depends on the degree or extent of division of labor shown in the economy of the animal. It is obvious that the same principle of division of labor with accompanying modification of structure is the basis of colonial and communal life. It is simply a manifestation of the principle among individuals instead of among organs. The division of the necessary labors of life among the different zooids of the colonial jelly-fish is plainly the reason for the profound and striking, but always reasonable and explicable modifications of the typical polyp or medusa body, which is shown by the swimming zooids, the feeding zooids, the sense zooids, and the others of the colony. And similarly in the case of the termite community, the soldier individuals are different structurally from the worker individuals because of the different work they have to do. And the queen differs from all the others, because of the extraordinary prolificacy demanded of her to maintain the great community.

It is important to note, however, that among those animals that show the most highly organized or specialized communal or social life, the structural differences among the individuals are the least marked, or at least are not the most profound. The three kinds of honey-bee individuals differ but little; indeed, as two of the kinds, male and female, are to be found in the case of almost all kinds of animals, whether communal in habit or not, the only unusual structural specialization in the case of the honey-bee, is the presence of the worker individual, which differs from the usual individuals in but little more than the rudimentary condition of the reproductive glands. Finally, in the case of man, with whom the communal or social habit is so all-important as to gain for him the name of "the social animal," there is no differentiation of individuals adapted only for certain kinds of work. Among these highest examples of social animals, the presence of an advanced mental endowment, the specialization of the mental power, the power of reason, have taken the place of and made unnecessary the structural differentiation of individuals. The honey-bee workers do different kinds of work: some gather food, some care for the young, and some make wax and build cells, but the individuals are interchangeable; each one knows enough to do these various things. There is a structural differentiation in the matter of only one special work or function, that of reproduction.

With the ants there is, in some cases, a considerable structural divergence among individuals, as in the genus *Atta* of South America with six kinds of individuals— namely, winged males, winged females, wingless soldiers, and wingless workers of three distinct sizes. In the case of other kinds with quite as highly organized a communal life there are but three kinds of individuals, the winged males and females and the wingless workers. The workers gather food, build the nest, guard the "cattle" (aphids), make war, and care for the young. Each one knows enough to do all these various distinct things. Its body is not so modified that it can do but one kind of thing, which thing it must always do.

The increase of intelligence, the development of the power of reasoning, is the most potent factor in the development of a highly specialized social life. Man is the example of the highest development of this sort in the animal kingdom, but the highest form of social development is not by any means the most perfectly communal.

268. **Advantages of communal life.**—The advantages of communal or social life, of co-operation and mutual aid, are real. The animals that have adopted such a life are among the most successful of all animals in the struggle for exist-

ence. The termite individual is one of the most defenseless, and, for those animals that prey on insects, one of
the most toothsome luxuries to be found in the insect
world. But the termite is one of the most abundant and
widespread and successfully living insect kinds in all the
tropics. Where ants are not, few insects are. The honey-
bee is a popular type of a successful life. The artificial
protection afforded the honey-bee by man may aid in its
struggle for existence, but it gains this protection because
of certain features of its communal life, and in Nature the
honey-bee takes care of itself well. The Little Bee People
of Kipling's Jungle Book, who live in great communities in
the rocks of Indian hills, can put to rout the largest and
fiercest of the jungle animals. Co-operation and mutual
aid are among the most important factors which help in
the struggle for existence. Its great advantages are, however, in some degree balanced by the fact that mutual help
brings mutual dependence. The community or society can
accomplish greater things than the solitary individuals, but
co-operation limits freedom, and often sacrifices the individual to the whole.

CHAPTER XXIII

COMMENSALISM AND PARASITISM

269. Association between animals of different species.—
The living together and mutual help discussed in the last
chapter concerned in each instance a single species of animal. All the various members of a pack of wolves or of a
community of ants are individuals of the same species.
But there are many instances of an association of individuals of different kinds of animals. In many cases of an
association of individuals of different species one kind
derives great benefit and the other suffers more or less
injury from the association. One kind lives at the expense
of the other. This association is called parasitism. In
some cases, however, neither kind of animal suffers from
the presence of the other. The two live together in harmony and presumably to their mutual advantage. In some
cases this mutual advantage is obvious. This kind of association is called *commensalism* or *symbiosis*.

270. Commensalism.—A curious example of commensalism is afforded by the different species of Remoras (*Echenei-
didæ*) which attach themselves to sharks, barracudas, and
other large fishes by means of a sucking disk on the top of
the head (Fig. 201). This disk is made by a modification
of the dorsal fin. The Remora thus attached to a shark
may be carried about for weeks, leaving its host only to
secure food. This is done by a sudden dash through the
water. The Remora injures the shark in no way save, perhaps, by the slight check its presence gives to the shark's
speed in swimming.

Whales, similarly, often carry barnacles about with them. These barnacles are permanently attached to the skin of the whale just as they would be to a stone or

Fig. 201.—Remora, with dorsal fin modified to be a sucking plate by which the fish attaches itself to a shark.

wooden pile. Many small crustaceans, annelids, mollusks, and other invertebrates burrow into the substance of living sponges for shelter. On the other hand, the little boring sponge (*Cliona*) burrows in the shells of oysters and other bivalves for protection. Some species of sponge "are never found growing except on the backs or legs of certain crabs." In these cases the sponge, with its many plant-like branches, protects the crab by concealing it from its enemies, while the sponge is benefited by being carried about by the crab to new food supplies.

Small fish of the genus *Nomeus* may often be found accompanying the beautiful Portuguese man-of-war (*Physalia*) as it sails slowly about on the ocean's surface (Fig. 202). These little fish lurk underneath the float and among the various hanging thread-like parts of the *Physalia*, which are provided with stinging cells.

In the nests of the various species of ants and termites many different kinds of other insects have been found. Some of these are harmful to their hosts, in that they feed on the food stores gathered by the industrious and provident ant, but others appear to feed only on refuse or useless substances in the nest. Some may even be of help to their hosts. Over one thousand species have been recorded by collectors as living habitually in the nests of ants and termites.

271. Symbiosis.—Of a more intimate character, and of more obvious and certain mutual advantage, is the well-known association called symbiosis. The hermit-crab always takes for his habitation the shell of another animal, often that of the common whelk. All of the hind part of the crab lies inside the shell, while its head with its great claws project from the opening of the shell. On the surface of the shell near the opening there is often to be found a sea-anemone, or sea-rose (Fig. 203). This sea-anemone is fastened securely to the shell, and has its mouth opening and tentacles near the head of the crab. The sea-anemone is carried from place to place by the hermit-crab, and in this way is much aided in obtaining food. On the other hand, the crab is protected from its enemies by the well-armed and dangerous tentacles of the sea-anemone. If the sea-anemone be torn away from the shell inhabited by one of these crabs, the crab will wander about, carefully seeking for another anemone. When he finds it he struggles to loosen it from its rock or

FIG. 202.—A Portuguese man-of-war (*Physalia*), with man-of-war fishes (*Nomeus gronovii*) living in the shelter of the stinging feelers. Specimens from off Tampa, Fla.

from whatever it may be growing on, and does not rest until he has torn it loose and placed it on his shell.

There are numerous small crabs called pea-crabs (*Pin-*

notheres) which live habitually inside the shells of living mussels. The mussels and the crabs live together in perfect harmony and to their mutual benefit.

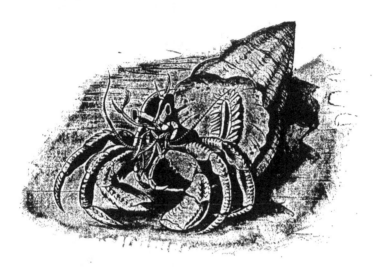

Fig. 203.—Hermit-crab (*Pagurus*) in shell, with a sea-anemone (*Adamsia palliata*) attached to the shell.—After Hertwig.

272. Relation of parasite and host.—There are many instances in the animal kingdom of an association between two animals by which one gains advantages great or small, sometimes even obtaining all the necessities of life, while the other gains nothing, but suffers corresponding disadvantage, often even the loss of life itself. This is the association between two animals whereby one, the parasite, lives on or in the other, the host, and at the expense of the host. Parasitism is a common phenomenon in all groups of animals; but parasites themselves are mostly invertebrates. When an animal can get along more safely or more easily by living at the expense of some other animal and takes up such a life, it becomes a parasite.

273. Kinds of parasitism.—The bird-lice (*Mallophaga*), which infest the bodies of all kinds of birds and are found

especially abundant on domestic fowls, live upon the outside of the bodies of their hosts, feeding upon the feathers
and dermal scales. They are examples of *external parasites.*
Other examples are fleas and ticks, and the crustaceans
called fish-lice and whale-lice, which are attached to marine
animals. On the other hand, almost all animals are infested
by certain parasitic worms which live in the alimentary
canal, like the tape-worm, or imbedded in the muscles, like
the trichina. These are examples of *internal parasites.*
Such parasites belong mostly to the class of worms, and
some of them are very injurious, sucking the blood from
the tissues of the host, while others feed solely on the partly
digested food. There are also parasites that live partly
within and partly on the outside of the body, like the *Sacculina*, which lives on various kinds of crabs. The body of
the *Sacculina* consists of a soft sac which lies on the outside
of the crab's body, and of a number of long, slender root-
like processes, which penetrate deeply into the crab's body,
and take up nourishment from within. The *Sacculina* is
itself a crustacean or crab-like creature. The classification
of parasites as external and internal is purely arbitrary, but
it is often a matter of convenience.

Some parasites live for their whole lifetime on or in the
body of the host, as is the case with the bird-lice. Their
eggs are laid on the feathers of the bird host; the young
when hatched remain on the bird during growth and development, and the adults only rarely leave the body, usually
never. These may be called *permanent parasites.* On the
other hand, fleas leap off or on a dog as caprice dictates;
or, as in other cases, the parasite may pass some definite
part of its life as a free, non-parasitic organism, attaching
itself, after development, to some animal, and remaining
there for the rest of its life. These parasites may be called
temporary parasites. But this grouping or classification,
like that of the external or internal parasites, is simply
a matter of convenience, and does not indicate at all

any blood relationship among the members of any one group.

274. The simple structure of parasites.—In all cases the body of a parasite is simpler in structure than the body of other animals which are closely related to the parasite— that is, animals that live parasitically have simpler bodies than animals that live free, active lives, competing for food with the other animals about them. This simplicity is not primitive, but results from the loss or atrophy of the structures which the mode of life renders useless. Many parasites are attached firmly to their host, and do not move about. They have no need of the power of locomotion. They are carried by their host. Such parasites are usually without wings, legs, or other locomotory organs. Because they have given up locomotion they have no need of organs of orientation, those special sense organs like eyes and ears and feelers, which serve to guide and direct the moving animal; and most non-locomotory parasites will be found to have no eyes, nor any of the organs of special sense which are accessory to locomotion and which serve for the detection of food or of enemies. Because these important organs, which depend for their successful activity on a highly organized nervous system, are lacking, the nervous system of parasites is usually very simple and undeveloped. Again, because the parasite usually has for its sustenance the already digested highly nutritious food elaborated by its host, most parasites have a very simple alimentary canal, or even no alimentary canal at all. Finally, as the fixed parasite leads a wholly sedentary and inactive life, the breaking down and rebuilding of tissue in its body go on very slowly and in minimum degree, and there is no need of highly developed respiratory and circulatory organs; so that most fixed parasites have these systems of organs in simple condition. They often bear no resemblance to the complex forms from which they are descended.

275. **Sacculina.**—Among the more highly organized animals the results of a parasitic life, in degree of structural
degeneration, can be more readily seen. A well-known parasite, belonging to the crustacea—the class of shrimps, crabs,
lobsters, and cray-fishes—is *Sacculina.* The young *Sacculina* is an active, free-swimming larva much like a young
prawn or young crab. But the adult bears absolutely no
resemblance to
such a typical
crustacean as a
cray-fish or crab.
The *Sacculina,*
after a short
period of inde-
pendent exist-
ence, attaches
itself to the ab-
domen of a crab,
and there com-
pletes its devel-
opment while

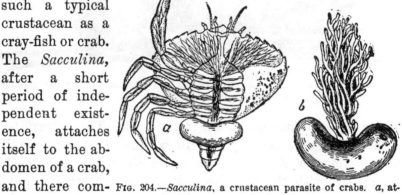

Fig. 204.—*Sacculina,* a crustacean parasite of crabs. *a,* attached to a crab, with root-like processes penetrating the
crab's body ; *b,* removed from the crab.

living as a parasite. In its adult condition (Fig. 204) it is
simply a great tumor-like sac, bearing many delicate, root-
like suckers, which penetrate the body of the crab host and
absorb nutriment. The *Sacculina* has no eyes, no mouth
parts, no legs, or other appendages, and hardly any of the
usual organs except reproductive organs. Degeneration here
is carried very far.

276. **Parasitic insects.**—In the order Hymenoptera there
are several families, all of whose members live during their
larval stage as parasites. We may call all these hymen-
opterons parasites ichneumon flies. The ichneumon flies
are parasites of other insects, especially of the larvæ of
beetles and moths and butterflies. In fact, the ichneumon
flies do more to keep in check the increase of injurious and
destructive caterpillars than do all our artificial remedies

for these insect pests. The adult ichneumon fly is four-
winged and lives an active, independent life. It lays its
eggs either in or on or near some caterpillar or beetle grub,
and the young ichneumon, when hatched, burrows about in
the body of its host, feeding on its tissues, but not attacking
such organs as the heart or nervous ganglia, whose injury

Fig. 205.—Parasitized caterpillar from which the ichneumon fly parasites have
issued, showing the circular holes of exit in the skin.

would mean immediate death to the host. The caterpillar
lives with the ichneumon grub within it, usually until
nearly time for its pupation. In many instances, indeed,
it pupates, with the parasite still feeding within its body,
but it never comes to maturity. The larval ichneumon fly
pupates either within the body of its host (Fig. 205) or
in a tiny silken cocoon outside of its body. From the
cocoons the adult winged ichneumon flies emerge, and
after mating find another host on whose body to lay their
eggs.

One of the most interesting ichneumon flies is *Thalessa*
(Fig. 209), which has a remarkably long, slender, flexible
ovipositor, or egg-laying organ. An insect known as the
pigeon horn-tail (*Tremex columba*) (Fig. 207) deposits its
eggs, by means of a strong, piercing ovipositor, half an inch
deep in the trunk wood of growing trees. The young or
larval *Tremex* is a soft-bodied white grub, which bores
deeply into the trunk of the tree, filling up the burrow behind it with small chips. The *Thalessa* is a parasite of the
Tremex, and when a female *Thalessa* finds a tree infested
by *Tremex*, she selects a place which she judges is opposite

a *Tremex* burrow, and, elevating her long ovipositor in a loop over her back, with its tip on the bark of the tree (Fig. 208), she makes a derrick out of her body and proceeds with

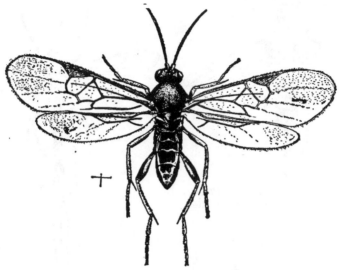

FIG. 206.—The adult ichneumon fly. The lines indicate natural dimensions.

great skill and precision to drill a hole into the tree. When the *Tremex* burrow is reached she deposits an egg in it. The larva that hatches from this egg creeps along this burrow until it reaches its victim, and then fastens itself to the horn-tail larva, which it destroys by sucking its blood. The larva of *Thalessa*, when full grown, changes to a pupa within the burrow of its host, and the adult gnaws a hole out through the bark if it does not find the hole already made by the *Tremex*.

277. **Degeneration through quiescence.**—If for any other reason animals should become fixed, and live inactive or sedentary lives, they would degenerate. And there are not a few instances of degeneration due simply to a quiescent life, unaccompanied by parasitism. The Tunicata, or sea-squirts (Fig. 210), are animals which have become simple through degeneration, due to the adoption of a sedentary

life, the withdrawal from the crowd of animals and from the struggle which it necessitates. The young tunicate is

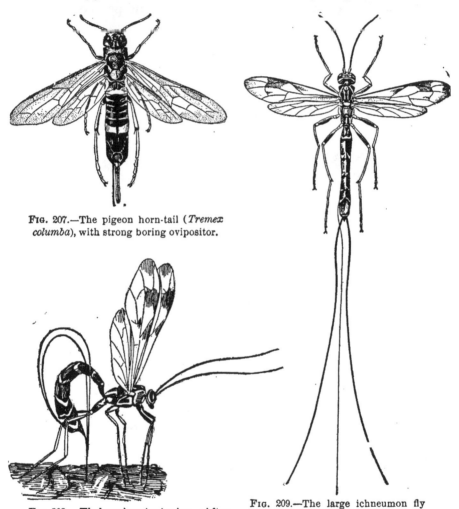

FIG. 207.—The pigeon horn-tail (*Tremex columba*), with strong boring ovipositor.

FIG. 208.—*Thalessa lunator* boring.—After COMSTOCK.

FIG. 209.—The large ichneumon fly *Thalessa*, with long flexible ovipositor. The various parts of this ovipositor are spread apart in the figure; naturally they lie together to form a single piercing organ.

a free-swimming, active, tadpole-like or fish-like creature, which possesses organs very like those of the adult of the simplest fishes or fish-like forms. That is, the sea-squirt begins life as a primitively simple vertebrate. It possesses in its larval stage a notochord, the delicate structure which precedes the formation of a backbone, extending along the upper part of the body, below the spinal cord. It is found in all young vertebrates, and is characteristic of the class. The other organs of the young tunicate are all of vertebral type. But the young sea-squirt passes a period of active and free life as a little fish, after which it settles down and attaches itself to a stone or shell or wooden pier by means of suckers, and remains for the rest of its life fixed. Instead of going on and developing into a fish-like creature, it loses its notochord, its special sense organs, and other organs; it loses its complexity and high organization, and becomes a "mere rooted bag with a double neck," a thoroughly degenerate animal.

Fig. 210.—A sea-squirt, or tunicate.

A barnacle is another example of degeneration through quiescence. The barnacles are crustaceans related most nearly to the crabs and shrimps. The young barnacle just from the egg (Fig. 211, f) is a six-legged, free-swimming nauplius, very like a young prawn or crab, with single eye. In its next larval stage it has six pairs of swimming feet, two compound eyes, and two large antennæ or feelers, and still lives an independent, free-swimming life. When it makes its final change to the adult condition, it attaches itself to some stone or shell, or pile or ship's bottom, loses its compound eyes and feelers, develops a protecting shell, and gives up all power of locomotion. Its swimming feet become changed into grasping organs, and it loses most of its outward resemblances to the other members of its class (Fig. 211, *e*).

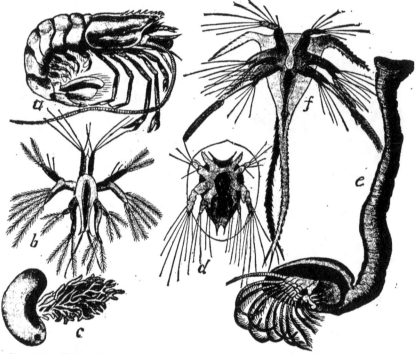

FIG. 211.—Three adult crustaceans and their larvæ. *a*, prawn (*Peneus*), active and free-living; *b*, larva of prawn; *c*, *Sacculina*, parasite; *d*, larva of *Sacculina*; *e*, barnacle (*Lepas*), with fixed quiescent life; *f*, larva of barnacle.—After HAECKEL.

Certain insects live sedentary or fixed lives. All the members of the family of scale insects (Coccidæ), in one sex at least, show degeneration, that has been caused by quiescence. One of these coccids, called the red orange scale (Fig. 211), is very abundant in Florida and California and in other orange-growing regions. The male is a beautiful, tiny, two-winged midge, but the female is a wingless, footless little sac without eyes or other organs of special sense, which lies motionless under a flat, thin, circular, reddish scale composed of wax and two or three cast skins of the insect itself.   The insect has a long, slender, flexible, sucking beak, which is thrust into the leaf or stem or fruit of the orange on which the "scale bug" lives and through which the insect sucks the orange sap, which is its only

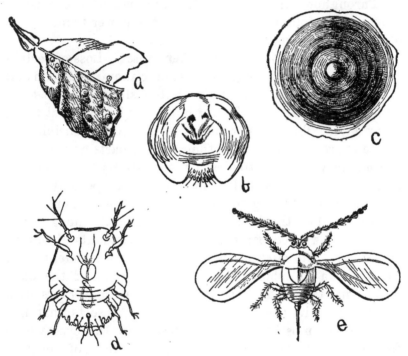

FIG. 212.—The red orange scale of California. *a*, bit of leaf with scales; *b*, adult female; *c*, wax scale under which adult female lives; *d*, larva; *e*, adult male.

food.   It lays eggs under its body, and thus also under the protecting wax scale, and dies.   From the eggs hatch active little larval scale-bugs with eyes and feelers and six legs. They crawl from under the wax scale and roam about over the orange tree.   Finally, they settle down, thrusting their sucking beak into the plant tissues, and cast their skin. The females lose at this molt their legs and eyes and feelers. Each becomes a mere motionless sac capable only of sucking up sap and of laying eggs. The young males, however, lose their sucking beak and can no longer take food, but they gain a pair of wings and an additional pair of eyes. They fly about and fertilize the sac-like females, which then molt again and secrete the thin wax scale over them.

Throughout the animal kingdom loss of the need of movement is followed by the loss of the power to move, and of all structures related to it.

278. Degeneration through other causes.—Loss of certain organs may occur through other causes than parasitism and a fixed life. Many insects live but a short time in their adult stage. May-flies live for but a few hours or, at most, a few days. They do not need to take food to sustain life for so short a time, and so their mouth parts have become rudimentary and functionless or are entirely lost. This is true of some moths and numerous other specially short-lived insects. Among the social insects the workers of the termites and of the true ants are wingless, although they are born of winged parents, and are descendants of winged ancestors. The modification of structure dependent upon the division of labor among the individuals of the community has taken the form, in the case of the workers, of a degeneration in the loss of the wings. Insects that live in caves are mostly blind; they have lost the eyes, whose function could not be exercised in the darkness of the cave. Certain island-inhabiting insects have lost their wings, flight being attended with too much danger. The strong sea-breezes may at any time carry a flying insect off the small island to sea. Only those which do not fly much survive, and by natural selection wingless breeds or species are produced. Finally, we may mention the great modifications of structure, often resulting in the loss of certain organs, which take place to produce protective resemblances (see Chapter XXIV). In such cases the body may be modified in color and shape so as to resemble some part of the environment, and thus the animal may be unperceived by its enemies. Many insects have lost their wings through this cause.

279. Immediate causes of degeneration.—When we say that a parasitic or quiescent mode of life leads to or causes degeneration, we have explained the stimulus or the ultimate cause of degenerative changes, but we have not shown just how parasitism or quiescence actually produces these changes. Degeneration or the atrophy and disappearance of organs or parts of a body is often said to be due to disuse. That is, the disuse of a part is believed by many naturalists to be the sufficient cause for its gradual dwindling and final loss. That disuse can so affect parts of a body during the lifetime of an individual is true. A muscle unused becomes soft and flabby and small. Whether the effects of such disuse can be inherited, however, is open to serious doubt. If not, some other immediate cause, or some other cause along with disuse, must be found. Such a cause must be sought for in the action of natural selection, preserving the advantages of simplicity of structure where action is not required.

CHAPTER XXIV

PROTECTIVE RESEMBLANCES, AND MIMICRY

280. Protective resemblance defined.—If a grasshopper be startled from the ground, you may watch it and determine exactly where it alights after its leap or flight, and yet, on going to the spot, be wholly unable to find it. Tho colors and marking of the insect so harmonize with its surroundings of soil and vegetation that it is nearly indistinguishable as long as it remains at rest. And if you were intent on capturing grasshoppers for fish-bait, this resemblance in appearance to their surroundings would be very annoying to you, while it would be a great advantage to the grasshoppers, protecting some of them from capture and death. This is *protective resemblance*. Mere casual observation reveals to us that such instances of protective resemblance are very common among animals. A rabbit or grouse crouching close to the ground and remaining motionless is almost indistinguishable. Green caterpillars lying outstretched along green grass-blades or on green leaves may be touched before being recognized by sight. In arctic regions of perpetual snow the polar bears, the snowy arctic foxes, and the hares are all pure white instead of brown and red and gray like their cousins of temperate and warm regions. Animals of the desert are almost without exception obscurely mottled with gray and sand color, so as to harmonize with their surroundings.

In the struggle for existence anything that may give an animal an advantage, however slight, may be sufficient to turn the scale in favor of the organism possessing the

advantage. Such an advantage may be swiftness of movement, or unusual strength or capacity to withstand unfavorable meteorological conditions, or the possession of such
color and markings or peculiar shape as tend to conceal the
animal from its enemies or from its prey. Resemblances
may serve the purpose of aggression as well as protection.
In the case of the polar bears and other predaceous animals that show color likenesses to their surroundings, the
resemblance can better be called aggressive than protective.
The concealment afforded by the resemblance allows them
to steal unperceived on their prey. This, of course, is an
advantage to them as truly as escape from enemies would be.

We have already seen that by the action of natural
selection and heredity those variations or conditions that
give animals advantages in the struggle for life are preserved and emphasized. And so it has come about that
advantageous protective resemblances are very widespread
among animals, and assume in many cases extraordinarily
striking and interesting forms. In fact, the explanation
of much of the coloring and patterning of animals depends
on this principle of protective resemblance.

Before considering further the general conditions of
protective resemblances, it will be advisable to refer to
specific examples classified roughly into groups or special
kinds of advantageous colorings and markings.

281. **General protective or aggressive resemblance.**—As
examples of general protective resemblance—that is, a general color effect harmonizing with the usual surroundings
and tending to hide or render indistinguishable the animal
—may be mentioned the hue of the green parrots of the
evergreen tropical forests; of the green tree-frogs and tree-
snakes which live habitually in the green foliage; of the
mottled gray and tawny lizards, birds, and small mammals of the deserts; and of the white hares and foxes
and snowy owls and ptarmigans of the snow-covered arctic regions. Of the same nature is the slaty blue of the

gulls and terns, colored like the sea. In the brooks most fishes are dark olive or greenish above and white below. To the birds and other enemies which look down on them from above they are colored like the bottom. To their fish enemies which look up from below, their color is like the white light above them, and their forms are not clearly seen. The fishes of the deep sea in perpetual darkness are

Fig. 213.—Alligator lizard (*Gerrhonotus scincicauda*) on granite rock. Photograph by J. O. Snyder, Stanford University, California.

inky violet in color below as well as above. Those that live among sea-weeds are red, grass-green, or olive, like the plants they frequent. General protective resemblance is very widespread among animals, and is not easily appreciated when the animal is seen in museums or zoölogical gardens—that is, away from its natural or normal environment. A modification of general color resemblance found in many animals may be called *variable* protective resemblance. Certain hares and other animals that live in northern latitudes are wholly white during the winter when the snow covers everything, but in summer, when much of the snow melts, revealing the brown and gray rocks and withered leaves, these creatures change color, putting on a grayish and brownish coat of hair.   The ptarmigan of the Rocky Mountains (one of the grouse), which lives on the snow and rocks of the high peaks, is almost wholly white in winter, but in summer when most of the snow is melted its plumage is chiefly brown.   On the campus at Stanford University there is a little pond whose shores are covered in some places with bits of bluish rock, in other places with bits of reddish rock, and in still other places with sand.   A small insect called the toad-bug (*Galgulus oculatus*) lives abundantly on the banks of this pond. Specimens collected from the blue rocks are bluish in color, those from the red rocks are reddish, and those from the sand are sand-colored.   Such changes of color to suit the changing surroundings can be quickly made in the case of some animals.   The chameleons of the tropics, whose skin changes color momentarily from green to brown, blackish or golden, is an excellent example of this highly specialized condition.   The same change is shown by a small lizard of our Southern States (*Anolius*), which from its habit is called the Florida chameleon.   There is a little fish (*Oligocottus snyderi*) which is common in the tide pools of the bay of Monterey, in California, whose color changes quickly to harmonize with the different colors of the rocks it happens to rest above.   Some of the tree-frogs show this variable coloring.   A very striking instance of variable protective resemblance is shown by the

Fig. 214.—Chrysalid of swallow-tail butterfly (*Papilio*), harmonizing with the bark on which it rests.

chrysalids of certain butterflies.   An eminent English naturalist collected many caterpillars of a certain species of butterfly, and put them, just as they were about to change into pupæ or chrysalids, into various boxes, lined with paper of different colors. The color of the chrysalid was found

Fig. 215.—Chrysalid of butterfly (lower left-hand projection from stem), showing protective resemblance. Photograph from Nature.

to harmonize very plainly with the color of the lining of the box in which the chrysalid hung. It is a familiar fact to entomologists that most butterfly chrysalids resemble in color and general external appearance the surface of the object on which they rest (Figs. 214 and 215).

282. **Special** protective **resemblance.**—Far more striking are those cases of protective resemblance in which the animal resembles in color and shape, sometimes in extraordinary detail, some particular object or part of its usual environment. Certain parts of the Atlantic Ocean are covered with great patches of sea-weed called the gulf-weed (*Sargassum*), and many kinds of animals—fishes and other creatures—live upon and among the algæ. No one can fail to note the extraordinary color resemblances which exist between those animals and the weed itself. The gulf-weed is of an olive-yellow color, and the crabs and shrimps, a certain flat-worm, a certain mollusk, and a little fish, all of which live among the *Sargassum*, are exactly of the same shade of yellow as the weed, and have small white markings on their bodies which are characteristic also of the *Sargassum*. The mouse-fish or *Sargassum* fish and the little sea-horses, often attached to the gulf-weed, show the same traits of coloration. In the black rocks about Tahiti is found the black noki or lava-fish (*Emmydrichthys vulcanus*) (Fig. 167), which corresponds perfectly in color and form to a piece of lava. This fish is also noteworthy for having envenomed spines in the fin on its back. The slender grass-green caterpillars of many moths and butterflies resemble very closely the thin grass-blades among which they live. The larvæ of the geometrid moths, called inch-worms or span-worms, are twig-like in appearance, and have the habit, when disturbed, of standing out stiffly from the twig or branch upon which they rest, so as to resemble in position as well as in color and markings a short or a broken twig. One of the most striking resemblances of this sort is shown by the large geometrid larva illustrated in Fig. 216, which was found near Ithaca, New York. The body of this caterpillar has a few small, irregular spots or humps, resembling very exactly the scars left by fallen

buds or twigs. These caterpillars have a special muscular development to enable them to hold themselves rigidly for

FIG. 216.—A geometrid larva on a branch. (The larva is the upper right-hand projection from the stem.)

FIG. 217.—A walking-stick insect (*Diapheromera femorata*) on twig.

long times in this trying attitude. They also lack the middle prop-legs of the body, common to other lepidopterous larvæ, the presence of which would tend to destroy the illusion so successfully carried out by them.   The common walking-stick (*Diapheromera*) (Fig. 217), with its wingless, greatly elongate, dull-colored body, is an excellent example of special protective resemblance.   It is quite indistinguishable, when at rest, from the twigs to which it is clinging. Another member of the family of insects to which the walking-stick belongs is the famous green-leaf insect (*Phyllium*)

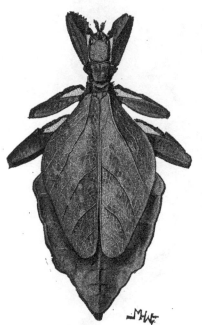

(Fig. 218).   It is found in South America and is of a bright green color, with broad leaf-like wings and body, with markings which imitate the leaf veins, and small irregular yellowish spots which mimic decaying or stained or fungus-covered spots in the leaf.

There are many butterflies that resemble dead leaves.   All our common meadow browns (*Grapta*), brown and reddish butterflies with ragged-edged wings, that appear in the autumn and flutter aimlessly about exactly like the falling leaves, show this resemblance.   But most remarkable of all is a

FIG. 218.—The green-leaf insect (*Phyllium*).

large butterfly (*Kallima*) (Fig. 219) of the East Indian region.   The upper sides of the wings are dark, with purplish and orange markings, not at all resembling a dead leaf.   But the butterflies when at rest hold their wings together over the back, so that only the under sides of the wings are exposed.   The under sides of *Kallima's* wings are exactly the color of a dead and dried leaf, and the wings are so held that all combine to mimic with extraordinary fidelity a dead leaf still attached to the twig by a short pedicle or leaf-stalk imitated by a short tail on the

FIG. 219.—*Kallima*, the "dead-leaf butterfly."

hind wings, and showing midrib, oblique veins, and, most remarkable of all, two apparent holes, like those made in leaves by insects, but in the butterfly imitated by two small circular spots free from scales and hence clear and transparent.   With the head and feelers concealed beneath the wings, it makes the resemblance wonderfully exact.

There are numerous instances of special protective resemblance among spiders.   Many spiders (Fig. 220) that

FIG. 220.—Spiders showing unusual shapes and patterns, for purposes of aggressive resemblance.

live habitually on tree trunks resemble bits of bark or small, irregular masses of lichen.   A whole family of spiders, which live in flower-cups lying in wait for insects, are white and pink and party-colored, resembling the markings of the special flowers frequented by them.   This is, of course, a

FIG. 221.—A pipe-fish (*Phyllopteryx*) resembling sea-weed, in which it lives.

special resemblance not so much for protection as for aggression ; the insects coming to visit the flowers are unable to distinguish the spiders and fall an easy prey to them.

283. **Warning colors and terrifying appearances.**—In the cases of advantageous coloring and patterning so far dis-

cussed the advantage to the animal lies in the resemblance between the animals and their surroundings, in the inconspicuousness and concealment afforded by the coloration. But there is another interesting phase of advantageous coloration in which the advantage derived is in rendering the animals as conspicuous and as readily recognizable as possible. While many animals are very inconspicuously colored, or are manifestly colored so as to resemble their surroundings, generally or specifically, many other animals are very brightly and conspicuously colored and patterned. If we are struck by the numerous cases of imitative coloring among insects, we must be no less impressed by the many cases of bizarre and conspicuous coloration among them. .

Many animals, as we well know, possess special and effective weapons of defense, as the poison-fangs of the venomous snakes and the stings of bees and wasps. Other animals, and with these cases most of us are not so well acquainted, possess a means of defense, or rather safety, in being inedible—that is, in possessing some acrid or ill-tasting substance in the body which renders them unpalatable to predaceous animals. Many caterpillars have been found, by observation in Nature and by experiment, to be distasteful to insectivorous birds. Now, it is obvious that it would be a great advantage to these caterpillars if they could be readily recognized by birds, for a severe stroke by a bird's bill is about as fatal to a caterpillar as being wholly eaten. Its soft, distended body suffers mortal hurt if cut or bitten by the bird's beak. This advantage of being readily recognizable is possessed by many if not all ill-tasting caterpillars by being brilliantly and conspicuously colored and marked. Such colors and markings are called warning colors. They are intended to inform birds of the fact that the caterpillar displaying them is an ill-tasting insect, a caterpillar to be let alone. The conspicuously black-and-yellow banded larva (Fig. 147, *b*) of the common

Monarch butterfly is a good example of the possession of warning colors by distasteful caterpillars.

These warning colors are possessed not only by the ill-tasting caterpillars, but by many animals which have special means of defense.   The wasps and bees, provided with stings—dangerous animals to trouble—are almost all conspicuously marked with yellow and black.   The lady-bird beetles (Fig. 222), composing a whole family of small beetles

FIG. 222.—Two lady-bird beetles, conspicuously colored and marked.

which are all ill-tasting, are brightly and conspicuously colored and spotted.   The Gila monster (*Heloderma*), the only poisonous lizard, differs from most other lizards in being strikingly patterned with black and brown.   Some of the venomous snakes are conspicuously colored, as the coral snakes (*Elaps*) or coralillos ot the tropics.   The naturalist Belt, whose observations in Nicaragua have added much to our knowledge of tropical animals, describes as follows an interesting example of warning colors in a species of frog: "In the woods around Santo Domingo (Nicaragua) there are many frogs.   Some are green or brown and imitate green or dead leaves, and live among foliage.   Others are dull earth-colored, and hide in holes or under logs.   All these come out only at night to feed, and they are all preyed upon by snakes and birds.   In contrast with these obscurely colored species, another little frog hops about in the daytime, dressed in a bright livery of red and blue. He can not be mistaken for any other, and his flaming breast and blue stockings show that he does not court concealment. He is very abundant in the damp woods, and I was convinced he was uneatable so soon as I made his acquaintance and saw the happy sense of security with which he hopped about. I took a few specimens home with me, and tried my fowls and ducks with them, but none would touch them. At last, by throwing down pieces of meat, for which there was a great competition among them, I managed to entice a young duck into snatching up one of the little frogs. Instead of swallowing it, however, it instantly threw it out of its mouth, and went about jerking its head, as if trying to throw off some unpleasant taste."

Certain animals which are without special means of defense and are not at all formidable or dangerous are yet so marked or shaped and so behave as to present a threatening or terrifying appearance. The large green caterpillars (Fig. 223) of the Sphinx moths—the tomato-worm is a familiar one of these larvæ—have a formidable-looking,

Fig. 223.—A "tomato-worm" larva of the Sphinx moth, *Phlegethontius carolina*, showing terrifying appearance.

sharp horn on the back of the next to last body ring. When disturbed they lift the hinder part of the body, bearing the horn, and move it about threateningly. As a matter of fact, the horn is not at all a weapon of defense, but is quite harmless. Numerous insects when disturbed lift the hind part of the body, and by making threatening motions lead enemies to believe that they possess a sting. The striking eye-spots of many insects are believed by some entomologists to be of the nature of terrifying appearances. The larva (Fig. 224) of the Puss moth (*Cerura*) has been often referred to as a striking example of terrifying appearances. When one of these larvæ is disturbed, "it retracts

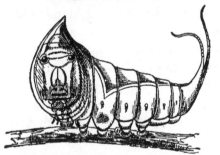

its head into the first body ring inflating the margin, which is of a bright red color. There are two intensely black spots on this margin in the appropriate position for eyes, and the whole appearance is that of a large flat face extending to the outer edge of the red margin. The effect is an intensely exaggerated caricature of a vertebrate face, which is probably alarming to the vertebrate enemies of the cat-

FIG. 224.—Larva of the Puss moth (*Cerura*). Upper figure shows the larva as it appears when undisturbed ; lower figure, when disturbed.—After POULTON.

erpillar. . . . The effect is also greatly strengthened by two pink whips which are swiftly protruded from the prongs of the fork in which the body terminates. . . . The end of the body is at the same time curved forward over the back, so that the pink filaments are brandished above the head."

284. **Alluring coloration.**—A few animals show what are called alluring colors—that is, they display a color pattern so arranged as to resemble or mimic a flower or other lure, and thus to entice to them other animals, their natural prey. This is a special kind of aggressive resemblance. A species

of predatory insect called a "praying-horse" (allied to the genus *Mantis*), found in India, has the shape and color of an orchid. Small insects are attracted and fall a prey to it. Certain Brazilian fly-catching birds have a brilliantly colored crest which can be displayed in the shape of a flower-cup. The insects attracted by the apparent flower furnish the fly-catcher with food. An Asiatic lizard is wholly colored like the sand upon which it lives except for a peculiar red fold of skin at each angle of the mouth. This fold is arranged in flower-like shape, "exactly resembling a little red flower which grows in the sand." Insects attracted by these flowers find out their mistake too late. In the tribe of fishes called the "anglers" or fishing frogs the front rays of the dorsal fin are prolonged in shape of long, slender fila-ments, the foremost and longest of which has a flattened and divided extremity like the bait on a hook. The fish conceals itself in the mud or in the cavities of a coral reef and waves the filaments back and forth. Small fish are at-tracted by the lure, mistaking it for worms writhing about in the water or among the weeds. As they approach they are ingulfed in the mouth of the angler, which in some of the species is of enormous size. One of these species is known to fishermen as the "all-mouth." These fishes (*Lophius piscatorius*), which live in the mud, are colored like mud or clay. Other forms of anglers, living among coral reefs, are brown and red (*Antennarius*), their colora-tion imitating in minutest detail the markings and out-growths on the reef itself, the lure itself imitating a worm of the reef. In a certain group of deep-sea anglers, the sea-devils (*Ceratiidæ*), certain species show a still further spe-cialization of the curious fishing-rod. In one species (*Co-rynolophus reinhardti*) (Fig. 156), living off the coast of Greenland at a depth of upward of a mile, the fishing-rod or first dorsal spine has a luminous bulb at its tip around which are fleshy, worm-like streamers. At the abyssal depths of a mile, more or less, frequented by these sea-devils there is no light, the inky darkness being absolute. This shining lure is therefore a most effective means of securing food.

285. **Mimicry.**—Although the word mimicry could often have been used aptly in the foregoing account of protective resemblances, it has been reserved for use in connection with a certain specific group of cases. It has been reserved to be applied exclusively to those rather numerous instances where an otherwise defenseless animal, one without poison-fangs or sting, and without an ill-tasting substance in its body, mimics some other specially defended or inedible animal sufficiently to be mistaken for it and so to escape attack. Such cases of protective resemblance are called true mimicry, and they are especially to be observed among insects.

In Fig. 225 are pictured three familiar American butterflies. One of these, the Monarch butterfly (*Anosia plexippus*), is perhaps the most abundant and widespread butterfly of our country. It is a fact well known to entomologists that the Monarch is distasteful to birds and is let alone by them. It is a conspicuous butterfly, being large and chiefly of a red-brown color. The Viceroy butterfly (*Basilarchia archippus*), also red-brown and much like the Monarch, is not, as its appearance would seem to indicate, a very near relative of the Monarch, belonging to the same genus, but on the contrary it belongs to the same genus with the third butterfly figured, the black and white *Basilarchia*. All the butterflies of the genus *Basilarchia* are black and white except this species, the Viceroy, and one other. The Viceroy is not distasteful to birds; it is edible, but it mimics the inedible Monarch so closely that the deception is not detected by the birds, and so it is not molested.

In the tropics there have been discovered numerous similar instances of mimicry by edible butterflies of inedible kinds. The members of two great families of butterflies (Danaidæ and Heliconidæ) are distasteful to birds, and are

FIG. 225.—The mimicking of the inedible Monarch butterfly by the edible Viceroy. Upper figure is the Monarch (*Anosia plexippus*); middle figure is the Viceroy (*Basilarchia archippus*); lowest figure is another member of the same genus (*Basilarchia*), to show the usual color pattern of the species of the genus.

mimicked by members of the other butterfly families (especially the Pieridæ), to which family our common white cabbage-butterfly belongs, and by the swallow-tails (Papilionidæ).

The bees and wasps are protected by their stings. They are usually conspicuous, being banded with yellow and black. They are mimicked by numerous other insects, especially moths and flies, two defenseless kinds of insects. This mimicking of bees and wasps by flies is very common, and can be observed readily at any flowering shrub. The flower-flies (Syrphidæ), which, with the bees, visit flowers, can be distinguished from the bees only by sharp observing. When these bees and flies can be caught and examined in hand, it will be found that the flies have but two wings while the bees have four.

A remarkable and interesting case of mimicry among insects of different orders is that of certain South American tree-hoppers (of the family Membracidæ, of the order Hemiptera), which mimic the famous leaf-cutting ant (*Sauba*) of the Amazons (Fig. 226). These ants have the curious habit of cutting off, with their sharp jaws, bits of green leaves and carrying them to their nests. In carrying the bits of leaves the ants hold them vertically above their heads. The leaf-hoppers mimic the ants and their burdens with remarkable exactitude by having the back of the body elevated in the form of a thin, jagged-edged ridge no thicker than a leaf. This part of the body is green like the leaves, while the under part of the body and the legs are brown like the ants.

Fig. 226.—Tree-hopper (Membracidæ), which mimics the leaf-cutting ant (*Sauba*) of Brazil. (Upper right-hand insect is the tree-hopper.)

Some examples of mimicry among other animals than

insects are known, but not many. The conspicuously marked venomous coral-snake or coralillos (*Elaps*) is mimicked by certain non-venomous snakes called king-snakes (*Lampropeltis, Oscoola*). The pattern of red and black bands surrounding the cylindrical body is perfectly imitated. But whether this is true mimicry brought about for purposes of protection may be doubted. Instances among birds have been described, and a single case has been recorded in the class of mammals. But it is among the insects that the best attested instances occur. The simple fact of the close resemblance of two widely related animals can not be taken to prove the existence of mimicry. Two animals may both come to resemble some particular part in their common environment and thus to resemble closely each other. Here we have simply two instances of special protective resemblance, and not an instance of mimicry. The student of zoölogy will do well to watch sharply for examples of protective resemblance or mimicry, for but few of the instances that undoubtedly exist are as yet known.

286. **Protective resemblances and mimicry most common among insects.**—The large majority of the preceding examples have been taken from among the insects. This is explained by the fact that the phenomena of protective resemblances and mimicry have been studied especially among insects; the theory of mimicry was worked out chiefly from the observation and study of the colors and markings of insects and of the economy of insect life. Why protective resemblances and mimicry among insects have been chiefly studied is because these conditions are specially common among insects. The great class Insecta includes more than two thirds of all the known living species of animals. The struggle for existence among the insects is especially severe and bitter. All kinds of "shifts for a living" are pushed to extremes; and as insect colors and patterns are especially varied and conspicuous, it is

only to be expected that this useful modification of colors and patterns, that results in the striking phenomena of special protective resemblances and mimicry, should be specially widespread and pronounced among insects. Moreover, they are mostly deficient in other means of defense, and seem to be the favorite food for many different kinds of animals. Protective resemblance is their best and most widely adopted means of preserving life.

287. No volition in mimicry.—The use of the word mimicry has been criticised because it suggests the exercise of volition or intent on the part of the mimicking animal. The student should not entertain this conception of mimicry. In the use of "mimicry" in connection with the phenomena just described, the biologist ascribes to it a technical meaning, which excludes any suggestion of volition or intent on the part of the mimic. Just how such extraordinary and perfect cases of mimicry as shown by *Phyllium* and *Kallima* have come to exist is a problem whose solution is not agreed on by naturalists, but none of them makes volition—the will or intent of the animal—any part of his proposed solution. Each case of mimicry is the result of a slow and gradual change, through a long series of ancestors. The mimicry may indeed include the adoption of certain habits of action which strengthen and make more pronounced the deception of shape and color. But these habits, too, are the result of a long development, and are instinctive or reflex—that is, performed without the exercise of volition or reason.

288. Color; its utility and beauty.—The causes of color, and the uses of color in animals and in plants are subjects to which naturalists have paid and are paying much attention. The subject of "protective resemblances and mimicry" is only one, though one of the most interesting, branches or subordinate subjects of the general theory of the uses of color. Other uses are obvious. Bright colors and markings may serve for the attraction of mates; thus are explained by some naturalists the brilliant plumage of the male birds, as in the case of the bird-of-paradise and the pheasants. Or they may serve for recognition characters, enabling the individuals of a band of animals readily to recognize their companions; the conspicuous whiteness of the short tail of the antelopes and cotton-tail rabbits, the black tail of the black-tail deer, and the white tail-feathers of the meadow-lark, are explained by many naturalists on this ground. Recognition marks of this type are especially numerous among the birds, hardly a species being without one or more of them, if their meaning is correctly interpreted. The white color of arctic animals may be useful not alone in rendering them inconspicuous, but may serve also a direct physiological function in preventing the loss of heat from the body by radiation. And the dark colors of animals may be of value to them in absorbing heat rays and thus helping them to keep warm. But "by far the most widespread use of color is to assist an animal in escaping from its enemies or in capturing its prey."

The colors of an animal may indeed not be useful to it at all. Many color patterns exist on present-day birds simply because, preserved by heredity, they are handed down by their ancestors, to whom, under different conditions of life, they may have been of direct use. For the most part, however, we can look on the varied colors and the striking patterns exhibited by animals as being in some way or another of real use and value. We can enjoy the exquisite coloration of the wings of a butterfly none the less, however, because we know that these beautiful colors and their arrangement tend to preserve the life of the dainty creature, and have been produced by the operation of fixed laws of Nature working through the ages.

CHAPTER XXV

THE SPECIAL SENSES

289. **Importance of the special senses.**—The means by which animals become acquainted with the outer world are the special senses, such as feeling, tasting, smelling, hearing, and seeing. The behavior of animals with regard to their surroundings, with regard to all the world outside of their own body, depends upon what they learn of this outer world through the exercise of these special senses. Habits are formed on the basis of experience or knowledge of the outer world gained by the special senses, and the development of the power to reason or to have sense depends on their pre-existence.

290. **Difficulty of the study of the special senses.**—We are accustomed to think of the organs of the special senses as extremely complex parts of the body, and this is certainly true in the case of the higher animals. In our own body the ears and eyes are organs of most specialized and highly developed condition. But we must not overlook the fact that the animal kingdom is composed of creatures of widely varying degrees of organization, and that in any consideration of matters common to all animals those animals of simplest and most lowly organization must be studied as well as those of high development. The study of the special senses presents two phases, namely, the study of the structure of the organs of special sense, and the study of the physiology of special sense—that is, the functions of these organs. It will be recognized that in the study of how other animals feel and taste and smell and hear and

371

see, we shall have to base all our study on our own experience. We know of hearing and seeing only by what we
know of our own hearing and seeing; but by examination
of the structure of the hearing and seeing organs of certain other animals, and by observation and experiments,
zoölogists are convinced that some animals hear sounds
that we can not hear, and some see colors that we can
not see.

While that phase of the study of the special senses
which concerns their structure may be quite successfully
undertaken, the physiological phase of the study of the
actual tasting and seeing and hearing of the lower animals
is a matter of much difficulty. The condition and character of the .special senses vary notably among different
animals. There may even exist other special senses than
the ones we possess. Some zoölogists believe that certain
marine animals possess a " density or pressure sense "—
that is, a sense which enables them to tell approximately
how deep in the water they may be at any time. To
certain animals is ascribed a " temperature sense," and
some zoölogists believe that what we call the homing instinct of animals as shown by the homing pigeons and
honey-bees and other animals, depends on their possession
of a special sense which man does not possess. Recent
experiments, however, seem to show that the homing of
pigeons depends on their keen sight. In numerous animals
there exist, besides the organs of the five special senses
which we possess, organs whose structure compels us to believe them to be organs of special sense, but whose function is wholly unknown to us. Thus in the study of the
special senses we are made to see plainly that we can not
rely simply on our knowledge of our own body structure
for an understanding of the structure and functions of
other animals.

291. **Special senses of the simplest animals.**—The *Amœba*,
described in Chapter I of Animal Life, is a one-celled

body, without organs, and yet with its capacity for performing the necessary life processes; there are no special senses except one (perhaps two). The *Amœba* can feel. It possesses the tactile sense. And there are no special sense organs except one, which is the whole of the outer surface of the body. If the *Amœba* be touched with a fine point it feels the touch, for the soft viscous protoplasm of its body flows slowly away from the foreign object. The sense of feeling or touch, the tactile sense, is the simplest or most primitive of the special senses, and the simplest, most primitive organ of special sense is the outer surface or skin of the body. Among those simple animals that possess the simplest organs of hearing and perceiving light, we shall find these organs to be simply specialized parts of the skin or outer cell layer of the body, and it is a fact that all the special sense organs of all animals are derived or developed from the outer cell layer, ectoblast, of the embryo. This is true also of the whole nervous system, the brain and spinal cord of the vertebrates, and the ganglia and nerve commissures of the invertebrates. And while in the higher animals the nervous system lies underneath the surface of the body, in many of the lower, many-celled animals all the ganglia and nerves, all of the nervous system, lie on the outer surface of the body, being simply a specialized part of the skin.

292. **The sense of touch.**—In some of the lower, many-celled animals, as among the polyps, there are on the skin certain sense cells, either isolated or in small groups, which seem to be stimulated not alone by the touching of foreign substances, but also by warmth and light. They are not limited to a single special sense. They are the primitive or generalized organs of special sense, and can develop into specialized organs for any one of the special senses.

The simplest and most widespread of these special senses with, as a whole, the simplest organs, is the tactile sense, or the sense of touch. The special organs of this sense are usually simple hairs or papillæ connecting with a nerve. These tactile hairs or papillæ may be distributed pretty evenly over most of the body, or may be mainly concentrated upon certain parts in crowded groups. Many of the lower animals have projecting parts, like the feeling tentacles of many marine invertebrates, or the antennæ (feelers) of crabs and insects, which are the special seat of the tactile organs. Among the vertebrates the tactile organs are either like those of the invertebrates, or are little sac-like bodies of connective tissue in which the end of a nerve is curiously folded and convoluted (Fig. 227). These little touch corpuscles simply lie in the cell layer of the skin, covered over thinly by the cuticle. Sometimes they are simply free, branched nerve-endings in the skin. These tactile corpuscles or free nerve-endings are especially abundant in those parts of the body which can be best used for feeling. In man the fin-ger-tips are thus especially supplied; in certain tailed monkeys the tip of the tail, and in hogs the end of the snout. The difference in abundance of these tactile corpuscles of the skin can be readily shown by experiment. With a pair of compasses, whose points have been slightly blunted, touch the skin of the forearm of a

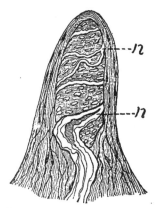

FIG. 227.—Tactile papilla of skin of man. n, nerve.— After KOELLIKER.

person who has his eyes shut, with the points about three inches apart and in the direction of the length of the arm. The person touched will feel the points as two. Repeat the touching several times, gradually lessening the distance between the points. When the points are not more than an inch to an inch and a half apart, the person touched will feel but a single touch—that is, the touching of both points will give the sensation of but a single contact. Repeat the experiment on the tip of the forefinger, and both points will be felt until the points are only about one tenth of an inch apart.

293. **The sense of taste.**—The sense of taste enables us to test in some degree the chemical constitution of substances which are taken into the mouth as food. We discriminate by the taste organs between good food and bad, well-tasting and ill-tasting. These organs are, with us and the other air-breathing animals, located in the mouth or on the mouth parts. They must be located so as to come into contact with the food, and it is also necessary that the food substance to be tasted be made liquid. This is accomplished by the fluids poured into the mouth from the salivary glands. With the lower aquatic animals it is not improbable that taste organs are situated on other parts of the body besides the mouth, and that taste is used not only to test food substances, but also to test the chemical character of the fluid medium in which they live.

The taste organs are much like the tactile organs, except that the special taste cell is exposed, so that small particles of the substance to be tasted can come into actual contact with it. The nerve-ending is usually in a small raised papilla or depressed pit. In the simplest animals there is no special organ of taste, and yet *Amœba* and other Protozoa show that they appreciate the chemical constitution of the liquid in which they lie. They taste—that is, test the chemical constitution of the substances—by means of their undifferentiated body surface. The taste organs are not always to be told from the organs of smell. Where an animal has a certain special seat of smell, like the nose of the higher animals, then the special sense organs of the mouth can be fairly assumed to be taste organs; but where the seat of both smell and taste is in the mouth or mouth parts, it is often impossible to distinguish between the two kinds of organs.

In mammals taste organs are situated on certain parts of the tongue, and have the form of rather large, low, broad papillæ, each bearing many small taste-buds (Fig. 228). In fishes similar papillæ and buds have been found in various ous places on the surface of the body, from which it is believed that the sense of taste in fishes is not limited to the mouth. In insects the taste-papillæ and taste-pits are grouped in certain places on the mouth parts, being especially abundant on

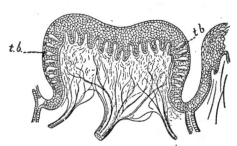

Fig. 228.—Vertical section of large papilla on tongue of a calf; *t. b.*, taste-buds. — After Lovén.

the tips of small, segmented, feeler-like processes called palpi, which project from the under lip and from the so-called maxillæ.

294. **The sense of smell.**—Smelling and tasting are closely allied, the one testing substances dissolved, the other testing substances vaporized. The organs of the sense of smell are, like those of taste, simple nerve-endings in papillæ or pits. The substance to be smelled must, however, be in a very finely divided form; it must come to the organs of smell as a gas or vapor, and not, as to the organs of taste, in liquid condition. The organs of smell are situated usually on the head, but as the sense of smell is used not alone for the testing of food, but for many other purposes, the organs of smell are not, like those of taste, situated principally in or near the mouth. Smell is a special sense of much wider range of use than taste. By smell animals can discover food, avoid enemies, and find their mates. They can test the air they breathe as well as the food they eat. In the matter of the testing of food the senses of both taste and smell are constantly used, and are indeed intimately associated.

The sense of smell varies a great deal in its degree of development in various animals. With the strictly aquatic animals—and these include most of the lower invertebrates, as the polyps, the star-fishes, sea-urchins, and most of the worms and mollusks—the sense of smell is probably but little developed. There is little opportunity for a gas or vapor to come to these animals, and only as a gas or vapor can a substance be smelled. With these animals the sense of taste must take the place of the olfactory sense. But among the insects, mostly terrestrial animals, there is an extraordinary development of the sense of smell. It is indeed probably their principal special sense. Insects must depend on smell far more than on sight or hearing for the discovery of food, for becoming aware of the presence of their enemies and of the proximity of their mates and companions. The organs of smell of insects are situated principally on the antennæ or feelers, a single pair of which is borne on the head of every insect (Fig. 229). That many insects have an extraordinarily keen sense of smell has been shown by numerous experiments, and is constantly proved by well-known habits. If a small bit of decaying flesh be inclosed in a box so that it is wholly concealed, it will nevertheless soon be found by the flies and carrion beetles that either feed on carrion or must always lay their eggs in decaying matter so that their carrion-eating larvæ may be provided with food. It is believed that ants find their way back to their nests by the sense of smell, and that they can recognize by scent among hundreds of individuals taken from various communities the members of their

FIG. 229.—Antenna of a leaf-eating beetle, showing smelling-pits on the expanded terminal segments.

own community. In the insectary at Cornell University, a few years ago, a few females of the beautiful promethea moth (*Callosamia promethea*) were inclosed in a box, which was kept inside the insectary building. No males had been seen about the insectary nor in its immediate vicinity, although they had been sought for by collectors. A few hours after the beginning of the captivity of the female moths there were forty male prometheas fluttering about over the glass roof of the insectary. They could not

Fig. 230.—Promethea moth, male, showing specialized antennæ.

see the females, and yet had discovered their presence in the building. The discovery was undoubtedly made by the sense of smell. These moths have very elaborately developed antennæ (Fig. 230), finely branched or feathered, affording opportunity for the existence of very many smelling-pits.

The keenness of scent of hounds and bird dogs is familiar to all, although ever a fresh source of astonishment as we watch these animals when hunting. We recently watched a retriever dog select unerringly, by the sense of smell, any particular duck out of a pile of a hundred. In the case of man the sense of smell is not nearly so well developed as among many of the other vertebrates. This inferiority is largely due to degeneration through lessened need; for in Indians and primitive races the sense of smell is keener and better developed than in civilized races. Where man has to make his living by hunting, and has to avoid his enemies of jungle and plain, his special senses are better developed than where the necessity of protection and advantage by means of such keenness of scent and hearing is done away with by the arts of civilization.

295. **The sense of hearing.**—Hearing is the perception of certain vibrations of bodies. These vibrations give rise to waves—sound waves as they are called—which proceed from the vibrating body in all directions, and which, coming to an animal, stimulate the special auditory or hearing organs, that transmit this stimulation along the auditory nerve to the brain, where it is translated as sound. These sound waves come to animals usually through the air, or, in the case of aquatic animals, through water, or through both air and water.

The organs of hearing are of very complex structure in the case of man and the higher vertebrates. Our ears, which are adapted for perceiving or being stimulated by vibrations ranging from 16 to 40,000 a second—that is, for hearing all those sounds produced by vibrations of a rapidity not less than 16 to a second nor greater than 40,000 to a second—are of such complexity of structure that many pages would be required for their description. But among the lower or less highly organized animals the ears, or auditory organs, are much simpler.

In most animals the auditory organs show the common characteristic of being wholly composed of, or having as an essential part, a small sac filled with liquid in which one or more tiny spherical hard bodies called *otoliths* are held. This auditory sac is formed of or lined internally by auditory cells, specialized nerve cells, which often bear delicate vibratile hairs (Fig. 231). Auditory organs of this general character are known among the polyps, the worms, the crustaceans, and the mollusks. In the common cray-fish the "ears" are situated in the basal segment of the inner antennæ or feelers (Fig. 232). They consist each of a small sac filled with liquid in which are suspended several grains of sand or other hard bodies. The inner

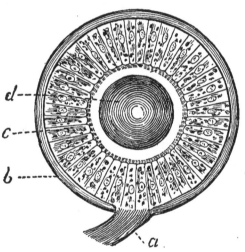

FIG. 231.—Auditory organ of a mollusk. *a*, auditory nerve; *b*, outer wall of connective tissue; *c*, cells with auditory hairs; *d*, otolith.—After LEYDIG.

FIG. 232. — Antenna of cray - fish, with auditory sac at base.— After HUXLEY.

surface of the sac is lined with fine auditory hairs. The sound waves coming through the air or water outside strike against this sac, which lies in a hollow on the upper or outer side of the antennæ. The sound waves are taken up by the contents of the sac and stimulate the fine hairs, which in turn give this stimulus to the nerves which run from them to the principal auditory nerve and thus to the brain of the cray-fish. Among the insects other kinds of auditory organs exist. The common locust or grasshopper has on the upper surface of the first abdominal segment a pair of tympana or ear-drums (Fig. 233), composed simply of the thinned, tightly stretched chitinous cuticle of the body.   On the inner surface of this

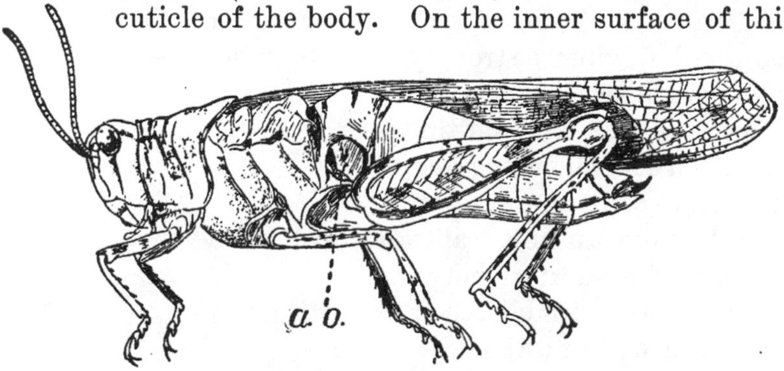

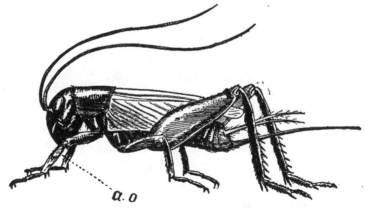

FIG. 233.—Grasshopper, showing auditory organ (*a. o.*) in first segment of abdomen. (Wings of one side removed.)

ear-drum there are a tiny auditory sac, a fine nerve leading from it to a small auditory ganglion lying near the tympanum, and a large nerve leading from this ganglion to one of the larger ganglia situated on the floor of the

FIG. 234.—A cricket, showing auditory organ (*a. o.*) in fore-leg.

thorax.   In the crickets and katydids, insects related to the locusts, the auditory organs or ears are situated in the fore-legs (Fig. 234).

Certain other insects, as the mosquitoes and other midges

or gnats, undoubtedly hear by means of numerous delicate hairs borne on the antennæ. The male mosquitoes (Fig. 235) have many hundreds of these long, fine antennal hairs, and on the sounding of a tuning-fork these hairs have been observed to vibrate strongly. In the base of each antenna there is a most elaborate organ, composed of fine chitinous rods, and accompanying nerves and nerve cells whose function it is to take up and transmit through the auditory nerve to the brain the stimuli received from the external auditory hairs.

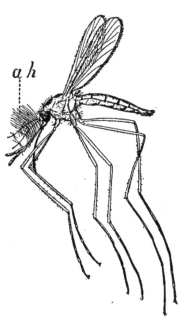

FIG. 235.—A male mosquito, showing auditory hairs (*a. h.*) on the antennæ.

296. **Sound - making.** — The sense of hearing enables animals not only to hear the warning natural sounds of storms and falling trees and plunging avalanches, but the sounds made by each other. Sound-making among animals serves to aid in frightening away enemies or in warning companions of their approach, for recognition among mates and members of a band or species, for the attracting and wooing of mates, and for the interchange of information. With the cries and roars of mammals, the songs of birds, and the shrilling and calling of insects all of us are familiar. These are all sounds that can be heard by the human ear. But that there are many sounds made by animals that we can not hear—that is, that are of too high a pitch for our hearing organs to be stimulated by—is believed by naturalists. Especially is this almost certainly true in the case of the insects. The peculiar sound-producing organs of

many sound-making insects are known; but certain other insects, that make no sound that we can hear, nevertheless possess similar sound-making organs.

Sound is produced by mammals and birds by the striking of the air which goes to and comes from the lungs against certain vibratory cords or flaps in the air-tubes. Sounds made by this vibration are re-enforced and made louder by arrangements of the air-tubes and mouth for resonance, and the character or quality of the sound is modified at will to a greater or less degree by the lips and teeth and other mouth structures. Sounds so made are said to be produced by a *voice*, or animals making sounds in this way are said to possess a voice. Animals possessing a voice have far more range and variety in their sound-making than most of the animals which produce sounds in other ways. The marvelous variety and the great strength of the singing of birds and of the cries and roars of mammals are unequaled by the sounds of any other animals.

But many animals without a voice—that is, which do not make sounds from the air-tubes—make sounds, and some of them, as certain insects, show much variety and range in their singing. The sounds of insects are made by the rapid vibrations of the wings, as the humming or buzzing of bees and flies, by the passage of air out or into the body through the many breathing pores or spiracles (a kind of voice), by the vibration of a stretched membrane or tympanum, as the loud shrilling of the cicada, and most commonly by stridulation—that is, by rubbing together two roughened parts of the body. The male crickets and the male katydids rub together the bases of their wing covers to produce their shrill singing. The locusts or grasshoppers make sounds when at rest by rubbing the roughened inside of their great leaping legs against the upper surface of their wing covers, and when in flight by striking the two wings of each side together. Numerous other insects make sounds by stridulation, but many of

these sounds are so feeble or so high in pitch that they are
rarely heard by us. Certain butterflies make an odd click-
ing sound, as do some of the water-beetles. In Japan,
where small things which are beautiful are prized not less
than large ones, singing insects are kept in cages and
highly valued, so that their capture becomes a lucrative
industry, just as it is with song birds in Europe and America. Among the many species of Japanese singing insects
is a night cricket, known as the bridle-bit insect, because
its note resembles the jingling of a bridle-bit.

297. **The sense of sight.**—Not all animals have eyes.
The moles which live underground, insects, and other animals that live in caves, and the deep-sea fishes which live
in waters so deep that the light of the sun never comes
to them, have no eyes at all, or have eyes of so rudimentary
a character that they can no longer be used for seeing.
But all these eyeless animals have no eyes because they
live under conditions where eyes are useless. They have
lost their eyes by degeneration. There are, however, many
animals that have no eyes, nor have they or their ancestors
ever had eyes. These are the simplest, most lowly organized animals. Many, perhaps all eyeless animals are, however, capable of distinguishing light from darkness. They
are sensitive to light. An investigator placed several individuals of the common, tiny fresh-water polyp (*Hydra*) in a
glass cylinder the walls of which were painted black. He
left a small part of the cylinder unpainted, and in this part
of the cylinder where the light penetrated the Hydras all
gathered. The eyeless maggots or larvæ of flies, when
placed in the light will wriggle and squirm away into dark
crevices. They are conscious of light when exposed to it,
and endeavor to shun it. Most plants turn their leaves
toward the light; the sunflowers turn on their stems to
face the sun. Light seems to stimulate organisms whether
they have eyes or not, and the organisms either try to get
into the light or to avoid it. But this is not seeing.

The simplest eyes, if we may call them eyes, are not capable of forming an image or picture of external objects. They only make the animal better capable of distinguishing between light and darkness or shadow. Many lowly organized animals, as some polyps, and worms, have certain cells of the skin specially provided with pigment. These cells grouped together form what is called a pigment fleck, which can, because of the presence of the pigment, absorb more light than the skin cells, and are more sensitive to the light. By such pigment-flecks, or eye-spots, the animal can detect, by their shadows, the passing near them of moving bodies, and thus be in some measure informed of the approach of enemies or of prey. Some of these eye-flecks are provided, not simply with pigment, but with a simple sort of lens that serves to concentrate rays of light and make this simplest sort of eye even more sensitive to changes in the intensity of light (Fig. 236).

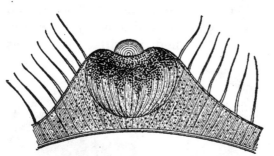

FIG. 236.—The simple eye of a jelly-fish (*Lizzia koellikeri*).—After O. and R. HERTWIG.

Most of the many-celled animals possess eyes by means of which a picture of external objects more or less nearly complete and perfect can be formed. There is great variety in the finer structure of these picture-forming eyes, but each consists essentially of an inner delicate or sensitive nervous surface called the *retina*, which is stimulated by light, and is connected with the brain by a large *optic nerve*, and of a transparent light-refracting lens lying outside of the retina and exposed to the light. These are the constant essential parts of an image-forming and image-perceiving eye. In most eyes there are other accessory parts which may make the whole eye an organ of excessively complicated structure and of remarkably perfect seeing capacity. Our own eyes are organs of extreme structural complexity and of high development, although some of the other vertebrates have undoubtedly a keener and more nearly perfected sight.

The crustaceans and insects have eyes of a peculiar character called compound eyes. In addition most insects have smaller simple eyes. Each of the compound eyes is composed of many (from a few, as in certain ants, to as many as twenty-five thousand, as in certain beetles) eye elements, each eye element seeing independently of the other eye elements and seeing only a very small part of any object in front of the whole eye. All these small parts of the external object seen by the many distinct eye elements are combined so as to form an image in mosaic—that is, made up of separate small parts—of the external object. If the head of a dragon-fly be examined, it will be seen that two thirds or more of the

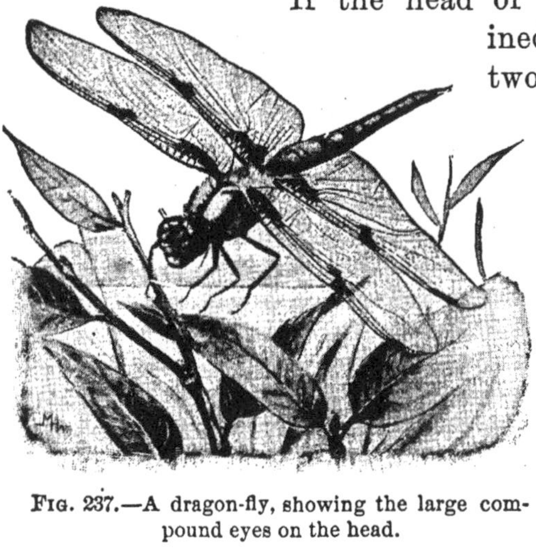

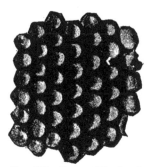

FIG. 237.—A dragon-fly, showing the large compound eyes on the head.

FIG. 238.—Some of the facets of the compound eye of a dragon-fly.

whole head is made up of the two large compound eyes (Fig. 237), and with a lens it may be seen that the outer surface of each of these eyes is composed of many small spaces or facets (Fig. 238) which are the outer lenses of the many eye elements composing the whole eye.

CHAPTER XXVI

INSTINCT AND REASON

298. **Irritability.**—All animals of whatever degree of organization show in life the quality of irritability or response to external stimulus. Contact with external things produces some effect on each of them, and this effect is something more than the mere mechanical effect on the matter of which the animal is composed. In the one-celled animals the functions of response to external stimulus are not localized. They are the property of any part of the protoplasm of the body. Just as breathing or digestion is a function of the whole cell, so are sensation and response in action. In the higher or many-celled animals each of these functions is specialized and localized. A certain set of cells is set apart for each function, and each organ or series of cells is released from all functions save its own.

299. **Nerve cells and fibers.**—In the development of the individual animal certain cells from the primitive external layer or ectoblast of the embryo are set apart to preside over the relations of the creature to its environment. These cells are highly specialized, and while some of them are highly sensitive, others are adapted for carrying or transmitting the stimuli received by the sensitive cells, and still others have the function of receiving sense-impressions and of translating them into impulses of motion. The nerve cells are receivers of impressions. These are gathered together in nerve masses or ganglia, the largest of these being known as the brain, the ganglia in general being known as nerve centers. The nerves are of two classes.

387

The one class, called sensory nerves, extends from the skin or other organ of sensation to the nerve center. The nerves of the other class, motor nerves, carry impulses to motion.

300. **The brain or sensorium.**—The brain or other nerve center sits in darkness surrounded by a bony protecting box. To this main nerve center, or *sensorium*, come the nerves from all parts of the body that have sensation, the external skin as well as the special organs of sight, hearing, taste, smell. With these come nerves bearing sensations of pain, temperature, muscular effort—all kinds of sensation which the brain can receive. These nerves are the sole sources of knowledge to any animal organism. Whatever idea its brain may contain must be built up through these nerve impressions. The aggregate of these impressions constitute the world as the organism knows it. All sensation is related to action. If an organism is not to act, it can not feel, and the intensity of its feeling is related to its power to act.

301. **Reflex action.**—These impressions brought to the brain by the sensory nerves represent in some degree the facts in the animal's environment. They teach something as to its food or its safety. The power of locomotion is characteristic of animals. If they move, their actions must depend on the indications carried to the nerve center from the outside; if they feed on living organisms, they must seek their food; if, as in many cases, other living organisms prey on them, they must bestir themselves to escape. The impulse of hunger on the one hand and of fear on the other are elemental. The sensorium receives an impression that food exists in a certain direction. At once an impulse to motion is sent out from it to the muscles necessary to move the body in that direction. In the higher animals these movements are more rapid and more exact. This is because organs of sense, muscles, nerve fibers, and nerve cells are all alike highly specialized. In the star-fish the sensation is slow, the muscular response sluggish, but the

method remains the same. This is simple reflex action, an impulse from the environment carried to the brain and then unconsciously reflected back as motion. The impulse of fear is of the same nature. Strike at a dog with a whip, and he will instinctively shrink away, perhaps with a cry. Perhaps he will leap at you, and you unconsciously will try to escape from him. Reflex action is in general unconscious, but with animals as with man it shades by degrees into conscious action, and into volition or action " done on purpose."

302. **Instinct.**—Different one-celled animals show differences in method or degree of response to external influences. The feelers of the *Amœba* will avoid contact with the feelers or pseudopodia of another *Amœba*, while it does not shrink from contact with itself or with an organism of unlike kind on which it may feed. Most Protozoa will discard grains of sand, crystals of acid, or other indigestible object. Such peculiarities of different forms of life constitute the basis of instinct.

Instinct is automatic obedience to the demands of external conditions. As these conditions vary with each kind of animal, so must the demands vary, and from this arises the great variety actually seen in the instincts of different animals. As the demands of life become complex, so may the instincts become so. The greater the stress of environment, the more perfect the automatism, for impulses to safe action are necessarily adequate to the duty they have to perform. If the instinct were inadequate, the species would have become extinct. The fact that its individuals persist shows that they are provided with the instincts necessary to that end. Instinct differs from other allied forms of response to external condition in being hereditary, continuous from generation to generation. This sufficiently distinguishes it from reason, but the line between instinct and reason and other forms of reflex action can not be sharply drawn.

It is not necessary to consider here the question of the origin of instincts. Some writers regard them as "inherited habits," while others, with apparent justice, doubt if mere habits or voluntary actions repeated till they become a "second nature" ever leave a trace upon heredity. Such investigators regard instinct as the natural survival of those methods of automatic response which were most useful to the life of the animal, the individuals having less effective methods of reflex action having perished, leaving no posterity.

An example in point would be the homing instinct of the fur-seal. When the arctic winter descends on its home in the Pribilof Islands in Bering Sea, these animals take to the open ocean, many of them swimming southward as far as the Santa Barbara Islands in California, more than three thousand miles from home. While on the long swim they never go on shore, but in the spring they return to the northward, finding the little islands hidden in the arctic fogs, often landing on the very spot from which they were driven by the ice six months before, and their arrival timed from year to year almost to the same day. The perfection of this homing instinct is vital to their life. If defective in any individual, he would be lost to the herd and would leave no descendants. Those who return become the parents of the herd. As to the others the rough sea tells no tales. We know that, of those that set forth, a large percentage never comes back. To those that return the homing instinct has proved adequate. This must be so so long as the race exists. The failure of instinct would mean the extinction of the species.

303. **Classification of instincts.**—The instincts of animals may be roughly classified as to their relation to the individual into egoistic and altruistic instincts.

Egoistic instincts are those which concern chiefly the individual animal itself. To this class belong the instincts of feeding, those of self-defense and of strife, the instincts of play, the climatic instincts, and environmental instincts, those which direct the animal's mode of life.

Altruistic instincts are those which relate to parenthood and those which are concerned with the mass of individuals of the same species. The latter may be called the social instincts. In the former class, the instincts of parenthood, may be included the instincts of courtship, reproduction, home-making, nest-building, and care for the young.

304. **Feeding.**—The instincts of feeding are primitively simple, growing complex through complex conditions. The protozoan absorbs smaller creatures which contain nutriment. The sea-anemone closes its tentacles over its prey. The barnacle waves its feelers to bring edible creatures within its mouth. The fish seizes its prey by direct motion. The higher vertebrates in general do the same, but the conditions of life modify this simple action to a very great degree.

In general, animals decide by reflex actions what is suitable food, and by the same processes they reject poisons or unsuitable substances. The dog rejects an apple, while the horse rejects a piece of meat. Either will turn away from an offered stone. Almost all animals reject poisons instantly. Those who fail in this regard in a state of nature die and leave no descendants. The wild vetches or "loco-weeds" of the arid regions affect the nerve centers of animals and cause dizziness or death. The native ponies reject these instinctively. This may be because all ponies which have not this reflex dislike have been destroyed. The imported horse has no such instinct and is poisoned. Very few animals will eat any poisonous object with which their instincts are familiar, unless it be concealed from smell and taste.

In some cases, very elaborate instincts arise in connection with feeding habits. With the California woodpeckers (*Melanerpes formicivorus bairdii*) a large number of them together select a live-oak tree for their operations. They first bore its bark full of holes, each large enough to hold an acorn. Then into each hole an acorn is thrust (Figs. 162 and 163). Only one tree in several square miles may be selected, and when their work is finished all those interested go about their business elsewhere. At irregular intervals a dozen or so come back with much clamorous discussion to look at the tree. When the right time comes, they all return, open the acorns one by one, devouring apparently the substance of the nut, and probably also the grubs of beetles which have developed within. When the nuts are ripe, again they return to the same tree and the same process is repeated. In the tree figured this has been noticed each year since 1891.

305. **Self-defense.**—The instinct of self-defense is even more varied in its manifestations. It may show itself either in the impulse to make war on an intruder or in the desire to flee from its enemies. Among the flesh-eating mammals and birds fierceness of demeanor serves both for the securing of food and for protection against enemies. The stealthy movements of the lion, the skulking habits of the wolf, the sly selfishness of the fox, the blundering good-natured power of the bear, the greediness of the hyena, are all proverbial, and similar traits in the eagle, owl, hawk, and vulture are scarcely less matters of common observation.

Herbivorous animals, as a rule, make little direct resistance to their enemies, depending rather on swiftness of foot, or in some cases on simple insignificance. To the latter cause the abundance of mice and mouse-like rodents may be attributed, for all are the prey of carnivorous beasts and birds, and even snakes.

Even young animals of any species show great fear of their hereditary enemies. The nestlings in a nest of the American bittern when one week old showed no fear of man, but when two weeks old this fear was very manifest

(Figs. 239 and 240). Young mocking-birds will go into spasms at the sight of an owl or a cat, while they pay little attention to a dog or a hen. Monkeys that have never seen a snake show almost hysterical fear at first sight of one, and the same kind of feeling is common to most men. A monkey was allowed to open a paper bag which

FIG. 239.—Nestlings of the American bittern. Two of a brood of four birds one week old, at which age they showed no fear of man. Photograph by E. H. TABOR, Meridian, N. Y., May 31, 1898. (Permission of Macmillan Company, publishers of Bird-Lore.)

contained a live snake. He was staggered by the sight, but after a while went back and looked in again, to repeat the experience. Each wild animal has its special instinct of resistance or method of keeping off its enemies. The stamping of a sheep, the kicking of a horse, the running in a circle of a hare, and the skulking in a circle of some foxes, are examples of this sort of instinct.

306. **Play.**—The play instinct is developed in numerous animals. To this class belong the wrestlings and mimic fights of young dogs, bear cubs, seal pups, and young beasts generally. Cats and kittens play with mice. Squir-

FIG. 240.—Nestlings of the American bittern. The four members of the brood of which two are shown in Fig. 153, two weeks old, when they showed marked fear of man. Photograph by F. M. CHAPMAN, Meridian, N. Y., June 8, 1898. (Permission of Macmillan Company, publishers of Bird-Lore.)

rels play in the trees. Perhaps it is the play impulse which leads the shrike or butcher-bird to impale small birds and beetles on the thorns about its nest, a ghastly kind of ornament that seems to confer satisfaction on the bird itself. The talking of parrots and their imitations of the sounds they hear seem to be of the nature of play. The greater their superfluous energy the more they will talk. Much of the singing of birds, and the crying, calling, and howling of other animals, are mere play, although singing primarily belongs to the period of reproduction, and other calls and cries result from social instincts or from the instinct to care for the young.

307. **Climate.**—Climatic instincts are those which arise from the change of seasons. When the winter comes the fur-seal takes its long swim to the southward; the wild geese range themselves in wedge-shaped flocks and fly high and far, calling loudly as they go; the bobolinks straggle away one at a time, flying mostly in the night, and most of the smaller birds in cold countries move away toward the tropics. All these movements spring from the migratory instinct. Another climatic instinct leads the bear to hide in a cave or hollow tree, where he sleeps or hibernates till spring. In some cases the climatic instinct merges in the homing instinct and the instinct of reproduction. When the birds move north in the spring they sing, mate, and build their nests. The fur-seal goes home to rear its young. The bear exchanges its bed for its lair, and its first business after waking is to make ready to rear its young.

308. **Environment.**—Environmental instincts concern the creature's mode of life. Such are the burrowing instincts of certain rodents, the woodchucks, gophers, and the like. To enumerate the chief phases of such instincts would be difficult, for as all animals are related to their environment, this relation must show itself in characteristic instincts.

309. **Courtship.**—The instincts of courtship relate chiefly to the male, the female being more or less passive. Among many fishes the male struts before the female, spreading his fins, intensifying his pigmented colors through museular tension, and in such fashion as he can makes himself the preferred of the female. In the little brooks in spring male minnows can be found with warts on the nose or head,

with crimson pigment on the fins, or blue pigment on the back, or jet-black pigment all over the head, or with varied combinations of all these. Their instinct is to display all these to the best advantage, even though the conspicuous hues lead to their own destruction. Against this contingency Nature provides a superfluity of males.

Among the birds the male in spring is in very many species provided with an ornamental plumage which he sheds when the breeding season is over. The scarlet, crimson, orange, blue, black, and lustrous colors of birds are commonly seen only on the males in the breeding season, the young males and all males in the fall having the plain brown gray or streaky colors of the female. Among the singing birds it is chiefly the male that sings, and his voice and the instinct to use it are commonly lost when the young are hatched in the nest.

Among polygamous mammals the male is usually much larger than the female, and his courtship is often a struggle with other males for the possession of the female. Among the deer the male, armed with great horns, fight to the death for the possession of the female or for the mastery of the herd. The fur-seal has on an average a family of about thirty-two females, and for the control of his harem others are ready at all times to dispute the possession. But with monogamous animals like the true or hair seal or the fox, where a male mates with a single female, there is no such discrepancy in size and strength, and the warlike force of the male is spent on outside enemies, not on his own species.

310. **Reproduction.**—The movements of many migratory animals are mainly controlled by the impulse to reproduce. Some pelagic fishes, especially flying-fishes and fishes allied to the mackerel, swim long distances to a region favorable for a deposition of spawn. Some species are known only in the waters they make their breeding homes, the individuals being scattered through the wide seas at

other times. Many fresh-water fishes, as trout, suckers, etc., forsake the large streams in the spring, ascending the small brooks where they can rear their young in greater safety. Still others, known as anadromous fishes, feed and mature in the sea, but ascend the rivers as the impulse of reproduction grows strong. Among such species are the salmon, shad, alewife, sturgeon, and striped bass in American waters. The most noteworthy case of the anadromous instinct is found in the king salmon or quinnat of the Pacific coast. This great fish spawns in November. In the Columbia River it begins running in March and April, spending the whole summer in the ascent of the river without feeding. By autumn the individuals are greatly changed in appearance, discolored, worn, and distorted. On reaching the spawning beds, some of them a thousand miles from the sea, the female deposits her eggs in the gravel of some shallow brook. After they are fertilized both male and female drift tail foremost and helpless down the stream, none of them ever surviving to reach the sea. The same habits are found in other species of salmon of the Pacific, but in most cases the individuals of other species do not start so early or run so far. A few species of fishes, as the eel, reverse this order, feeding in the rivers and brackish creeks, dropping down to the sea to spawn.

The migration of birds has relation to reproduction as well as to changes of weather. As soon as they reach their summer homes, courtship, mating, nest-building, and the care of the young occupy the attention of every species.

311. **Care of the young.**—In the animal kingdom one of the great factors in development has been the care of the young. This feature is a prominent one in the specialization of birds and mammals. When the young are cared for the percentage of loss in the struggle for life is greatly reduced, the number of births necessary to the maintenance of the species is much less, and the opportunities for specialization in other relations of life are much greater.

In these regards, the nest-building and home-making animals have the advantage over those that have not these instincts. The animals that mate for life have the advantage over polygamous animals, and those whose social or mating habits give rise to a division of labor over those with instincts less highly specialized.

The interesting instincts and habits connected with nest or home building and the care of the young are discussed in the next chapter.

312. **Variability of instincts.**—When we study instincts of animals with care and in detail, we find that their regularity is much less than has been supposed. There is as much variation in regard to instinct among individuals as there is with ·regard to other characters of the species. Some power of choice is found in almost every operation of instinct. Even the most machine-like instinct shows some degree of adaptability to new conditions. On the other hand, in no animal does reason show entire freedom from automatism or reflex action. " The fundamental identity of instinct with intelligence," says an able investigator, " is shown in their dependence upon the same structural mechanism (the brain and nerves) and in their responsive adaptability."

313. **Reason.**—Reason or intellect, as distinguished from instinct, is the choice, more or less conscious, among responses to external impressions. Its basis, like that of instinct, is in reflex action. Its operations, often repeated, become similarly reflex by repetition, and are known as habit. A habit is a voluntary action repeated until it becomes reflex. It is essentially like instinct in all its manifestations. The only evident difference is in its origin. Instinct is inherited. Habit is the reaction produced within the individual by its own repeated actions. In the varied relations of life the pure reflex action becomes inadequate. The sensorium is offered a choice of responses. To choose one and to reject the others is the function of intel-

lect or reason. While its excessive development in man·
obscures its close relation to instinct, both shade off by
degrees into reflex action. Indeed, no sharp line can be
drawn between unconscious and subconscious choice of
reaction and ordinary intellectual processes.

Most animals have little self-consciousness, and their
reasoning powers at best are of a low order; but in kind,
at least, the powers are not different from reason in man.
A horse reaches over the fence to be company to another.
This is instinct. When it lets down the bars with its teeth,
that is reason. When a dog finds its way home at night by
the sense of smell, this may be instinct; when he drags a
stranger to his wounded master, that is reason. When a
jack-rabbit leaps over the brush to escape a dog, or runs in
a circle before a coyote, or when it lies flat in the grass as a
round ball of gray indistinguishable from grass, this is instinct. But the same animal is capable of reason—that is,
of a distinct choice among lines of action. Not long ago a
rabbit came bounding across the university campus at Palo
Alto. As it passed a corner it suddenly faced two hunting
dogs running side by side toward it. It had the choice of
turning back, its first instinct, but a dangerous one; of
leaping over the dogs, or of lying flat on the ground. It
chose none of these, and its choice was instantaneous. It
ceased leaping, ran low, and went between the dogs just as
they were in the act of seizing it, and the surprise of the
dogs, as they stopped and tried to hurry around, was the
same feeling that a man would have in like circumstances.

On the open plains of Merced County, California, the
jack-rabbit is the prey of the bald eagle. Not long since a
rabbit pursued by an eagle was seen to run among the
cattle. Leaping from cow to cow, he used these animals
as a shelter from the savage bird. When the pursuit was
closer, the rabbit broke cover for a barbed wire fence.
When the eagle swooped down on it, the rabbit moved a
few inches to the right, and the eagle could not reach him

through the fence. When the eagle came down on the other side, he moved across to the first. And this was continued until the eagle gave up the chase. It is instinct that leads the eagle to swoop on the rabbit. It is instinct again for the rabbit to run away. But to run along the line of a barbed wire fence demands some degree of reason. If the need to repeat it arose often in the lifetime of a single rabbit it would become a habit.

The difference between intellect and instinct in lower animals may be illustrated by the conduct of certain monkeys brought into relation with new experiences. At one time we had two adult monkeys, "Bob" and "Jocko," belonging to the genus *Macacus*. Neither of these possessed the egg-eating instinct. At the same time we had a baby monkey, "Mono," of the genus *Cercopithecus*. Mono had never seen an egg, but his inherited impulses bore a direct relation to feeding on eggs, just as the heredity of *Macacus* taught the others how to crack nuts or to peel fruit.

To each of these monkeys we gave an egg, the first that any of them had ever seen. The baby monkey, Mono, being of an egg-eating race, devoured his egg by the operation of instinct or inherited habit. On being given the egg for the first time, he cracked it against his upper teeth, making a hole in it, and sucked out all the substance. Then holding the egg-shell up to the light and seeing that there was no longer anything in it, he threw it away. All this he did mechanically, automatically, and it was just as well done with the first egg he ever saw as with any other he ate. All eggs since offered him he has treated in the same way.

The monkey Bob took the egg for some kind of nut. He broke it against his upper teeth and tried to pull off the shell, when the inside ran out and fell on the ground. He looked at it for a moment in bewilderment, took both hands and scooped up the yolk and the sand with which it was mixed and swallowed the whole. Then he stuffed the shell itself into his mouth. This act was not instinctive. It was the work of pure reason. Evidently his race was not familiar with the use of eggs and had acquired no instincts regarding them. He would do it better next time. Reason is an inefficient agent at first, a weak tool; but when it is trained it becomes an agent more valuable and more powerful than any instinct.

The monkey Jocko tried to eat the egg offered him in much the same way that Bob did, but, not liking the taste, he threw it away.

The confusion of highly perfected instinct with intellect is very common in popular discussions. Instinct grows weak and less accurate in its automatic obedience as the intellect becomes available in its place. Both intellect and instinct are outgrowths from the simple reflex response to external conditions. But instinct insures a single definite response to the corresponding stimulus. The intellect has a choice of responses. In its lower stages it is vacillating and ineffective; but as its development goes on it becomes alert and adequate to the varied conditions of life. It grows with the need for improvement. It will therefore become impossible for the complexity of life to outgrow the adequacy of man to adapt himself to its conditions.

Many animals currently believed to be of high intelligence are not so. The fur-seal, for example, finds it way back from the long swim of two or three thousand miles through a foggy and stormy sea, and is never too late or too early in arrival. The female fur-seal goes two hundred miles to her feeding grounds in summer, leaving the pup on the shore. After a week or two she returns to find him within a few rods of the rocks where she had left him. Both mother and young know each other by call and by odor, and neither is ever mistaken, though ten thousand other pups and other mothers occupy the same rookery. But this is not intelligence. It is simply instinct, because it has no element of choice in it. Whatever its ancestors were forced to do the fur-seal does to perfection. Its instincts are perfect as clockwork, and the necessities of migration must keep them so. But if brought into new conditions it is dazed and stupid. It can not choose when different lines of action are presented.

The Bering Sea Commission once made an experiment on the possibility of separating the young male fur-seals, or "killables," from the old ones in the same band. The method was to drive them through a wooden chute or runway with two valve-like doors at the end. These animals can be driven like sheep, but to sort them in the way proposed proved impossible. The most experienced males would beat their noses against a closed door, if they had seen a seal before them pass through it. That this door had been shut and another opened beside it passed their comprehension. They could not choose the new direction. In like manner a male fur-seal will watch the killing and skinning of his mates with perfect composure. He will sniff at their blood with languid curiosity; so long as it is not his own it does not matter. That his own blood may flow out on the ground in a minute or two he can not foresee.

Reason arises from the necessity for a choice among actions. It may arise as a clash among instincts which forces on the animal the necessity of choosing. A doe, for example, in a rich pasture has the instinct to feed. It hears the hounds and has the instinct to flee. Its fawn may be with her and it is her instinct to remain and protect it. This may be done in one of several ways. In proportion as the mother chooses wisely will be the fawn's chance of survival. Thus under difficult conditions, reason or choice among actions rises to the aid of the lower animals as well as man.

314. **Mind.**—The word mind is popularly used in two different senses. In the biological sense the mind is the collective name for the functions of the sensorium in men and animals. It is the sum total of all psychic changes,

actions and reactions. Under the head of psychic functions
are included all operations of the nervous system as well as
all functions of like nature which may exist in organisms
without specialized nerve fibers or nerve cells. As thus defined mind would include all phenomena of irritability, and
even plants have the rudiments of it. The operations of
the mind in this sense need not be conscious. With the
lower animals almost all of them are automatic and unconscious. With man most of them must be so. All functions of the sensorium, irritability, reflex action, instinct,
reason, volition, are alike in essential nature though differing greatly in their degree of specialization.

In another sense the term mind is applied only to conscious reasoning or conscious volition. In this sense it is
mainly an attribute of man, the lower animals showing it
in but slight degree. The discussion as to whether lower
animals have minds turns on the definition of mind, and
our answer to it depends on the definition we adopt.

A "pointer" dog in the act of "pointing," a specialized instinct.
(Permission of G. O. Shields, publisher of Recreation.)

CHAPTER XXVII

ECONOMIC ZOOLOGY

315. Uses of animals to man.—Economic zoology treats of the value of animals for the purposes of man. These services are enormously varied, and in this chapter we can give ouly a bare enumeration of some of the most conspicuous lines of service, leaving the student to develop the details. At the outset we may remember that most of the species of animals have inhabited the earth longer than man has, and that we have no right to suppose that the reason for their creation was to render him some service. Thousands and thousands of species can be of no possible use in human affairs, and a few are related to man only through their ability to inflict positive injury. Of harmful nature are the insects with poison glands connected with the mouth, many of those with stings, the snakes with venom fangs, the poisonous Gila monster among lizards, some of the great beasts of prey, and, perhaps most of all, the noxious types of mosquitoes, who transfer to the human body the germs of certain diseases, as malaria, yellow fever, and filariasis. Other noxious animals are the vermin—rats and mice and the like—which infest houses and may carry disease, the many forms of internal and external parasites, intestinal worms, ticks, mites, and the like. Harmful in other ways are the hordes of insects injurious to vegetation; and some mammals, as rabbits and gophers, are at times extremely destructive to valuable plants.

316. Domestic animals.—The very earliest records of man show that he trained those animals about him which could be made to minister to his needs or his pleasure. The young of almost any species can be rendered friendly and fearless by kind treatment. Naturally those most easily tamed and most useful when reared received the most attention. Of the young born in domestication, those most tractable or most helpful would be most cherished. Thus during the lapse of ages by a process of selection, conscious or unconscious, distinct breeds were formed, many of these differing from the original stock more than distinct species differ from each other. Varying needs brought greater and greater differences among breeds; thus all dogs are domesticated wolves of different species, but the distinctions between St. Bernard dogs, Eskimo dogs, greyhounds, pugs, lap-dogs, and the tiny hairless Pelon dogs of Mexico are far greater than the differences separating different kinds of wolves.·

317. **Formation of new races.**—With the advance of civilization unconscious selection has developed into conscious choice, and new and improved races can be planned and developed with almost absolute certainty of success. Selective breeding has been called "the magician's wand," by which the breeder can summon up new forms useful to man or pleasing to his fancy. In general, new varieties are formed by crossing old ones, each of which has certain desirable traits. These may be combined in certain of the young, or other qualities, new or unforeseen, may appear. These are retained as the basis of the improved race, while those individuals not possessing the desired characters are discarded. By pursuing this method for a certain number of generations the new type may become more or less perfectly established. In this regard almost any desired result is possible with time and patience. Those species most widely domesticated have, in general, developed the greatest number of distinct races or breeds. Among these are the dog, the horse, the donkey, the ox, the sheep, the goat, the hog (descended from the wild boar), the rabbit,

the cat, the fowl, the goose, the duck, the peacock, the guinea-hen, the camel, the honey-bees, the silk-worm, the elephant, the llama, the reindeer, the falcon, the turkey, the ferret, the different parrots, the guinea-pig, and other species. Those forms domesticated for special purposes or within a narrow range are less likely to form varieties.

318. Artificial propagation.—Many animals are bred within regions not formerly occupied by them, although being in no sense domesticated. Among these are the various kinds of salmon and trout, the shad, the striped bass, the carp, goldfish, and many other fishes, the oyster, the Chinese pheasant, the lady-birds. In some cases the eggs are taken and hatched under artificial conditions. This is especially the case with the salmon and trout. In other cases the animals are simply liberated in a new region to make their way in competition with other species.

319. Services of animals.—The chief services rendered by animals may be treated under the following heads:

Food, clothing, ornaments, use in the arts, as destroyers of injurious animals, as servants, and as friends.

320. Animals used as food.—All races of men have fed, in part at least, on the flesh of animals, either raw or cooked. For this purpose certain species have long been domesticated. As a rule, those mammals and birds which are wholly carnivorous have been rejected by man as unfit for food; but this rule does not apply to the class of fishes. Among the animals whose flesh is especially valued may be named the ox, the sheep, the goat tribe, the deer tribe, the hog, and, in general, all hoofed animals with four toes. Besides these, the various rabbits, squirrels, bears, raccoons, opossums, fur seals, the large bats, certain monkeys, some whales, and a variety of other mammals are largely eaten by men on account of the excellence of their flesh. All mammals, excepting the strictly carnivorous cats and wolves, are considered welcome food by some races of men, and even these have not been wholly rejected.

The milk of the larger hoofed animals—the cow, sheep, goat, buffalo, and even the horse and the ass—has formed an important part in human diet.

All the larger birds which are not strictly carnivorous are eaten by man, and the eggs of these and many others, domestic birds and wild birds, have formed a large part of his diet. In China a certain species of swallow (*Collocalia*) forms a nest in part from a secretion from its own stomach. This substance forms an agreeable basis for soup, the so-called edible bird's-nest.

Among the reptiles certain species of turtles have flesh of great delicacy—for instance, certain land species, as the Maryland diamond-back terrapin, and some of the great sea-turtles. Among the amphibians, the chief food product is found in the delicate muscles of the legs of various species of frogs.

Of the 12,000 known species of fishes, many are too small to be worth taking. These serve as food for larger fishes. A few dozen species in the tropics have flesh containing a bitter alkaloid, which is more or less poisonous. The great majority are, however, excellent as food, and upward of 5,000 species may be fairly called food-fishes. Certain fishes yield jelly-like substances from the air-bladder or other structures. The eggs of the sturgeon are prepared to be eaten as " caviar." Among the fishes most delicious as food, the eulachon of the Columbia River perhaps ranks first. The ayu, or samlet, of Japan resembles it. Next we may place the pámpano of the Gulf of Mexico, the Spanish mackerel of the same region, the whitefish of the Great Lakes, the bluefish and weakfish of New England, and the various kinds of trout, grayling, bass, shad, and pickerel. The sole, the surmullet, and the turbot rank among the first of the fishes of Europe. Of far greater economic value than any of these, from their exceeding abundance, are the various kinds of salmon, the cod, herring, mackerel, and halibut.

The gelatinous fin-rays of certain sharks (see Fig. 103) make an excellent soup, much valued by the Chinese.

In most regions the flesh of the fish is cooked before eating. In arctic regions it is salted or smoked. In Japan and Hawaii fishes are largely eaten raw.

Insects, such as locusts and the larvæ of certain beetles, are used as human foods by the lowest races only. The honéy made from nectar gathered from flowers by the honey

Fig. 241.—The palolo or edible worm of the Atlantic (*Eunice fucata Ehlers*), Tortugas, Florida.—After A. G. Mayer.

bee has, however, been regarded as a delicacy by all races of men who dwell in regions inhabited by bees. Various crustaceans, as the lobster, cray-fish, and many crabs, are, however, much esteemed. The "*bêche de mer*" is a holothurian used as food in the western Pacific; and many people eat certain species of sea-urchins. Worms have rarely any economic value. The common earthworm is, however, of the greatest service in pulverizing soils. Certain sea-worms are edible. Most notable of these is a worm of the coral reefs called palolo, found about Samoa and Fiji, and a second species in the tropical Atlantic. Twice during the year—in October and November—the posterior half of the body, bearing the reproductive elements, separates from the head end of the animal, and swims to the surface for spawning purposes. This phenomenon occurs at definite times—at dawn of the day on which the moon is in its last quarter, and on the day previous. At this season the water fairly boils with countless thousands of these headless worms, that are collected by the natives and esteemed a great delicacy.

Among the mollusks many species are excellent as food for man. Foremost among these is the oyster. With it are many species of clam, scallop, cockle, snail, abalone, squid, cuttlefish, and octopus.

321. **Clothing from animals.**—The hair of certain mammals may be used as a fabric for cloth. The most valuable in this connection is the wool of sheep. Wild sheep have little or no wool, the great yield of this article being a product of artificial selection. A coarser wool is produced by some goats, as also by some animals related to the Peruvian llama. The hair of the camel is used to make a coarse cloth.

Another textile fabric of great importance is silk, which is the fine-spun covering of the larvæ and chrysalids of a white moth. The furs and skins of many animals formed the chief clothing of primitive man, and are still largely used in cold regions among fashionable as well as primitive people. Among the finest of furs are those of the weasel tribe, the otter, mink, ermine, marten, and their relatives. Most valuable of all these is the fur of the sea-otter of the north Pacific, a single skin being sometimes valued at upward of $1,000. The female and young of the fur seal (*Otoes*) yield a very soft and beautiful fur after the long hairs have been plucked out. Many foxes yield delicate and beautiful furs. In general, those animals living in the coldest regions are most valued, because their hairs are longer and more closely set than in similar animals living with less need of protection from the cold. Coarser furs are taken from the various bears, wolves, and other carnivorous animals, from wild goats and other animals of northern or mountainous regions. The skins of tigers, leopards, and other large members of the cat family, with short, close-set, glossy fur, often beautifully colored, are also highly valued. The long, coarse hair of the buffalo caused the "buffalo-robe" to be highly appreciated by our fathers before the species was exterminated. The skins of animals with indifferent fur, as squirrels and prairie-dogs, are often stitched together to form blankets.

The skins of many animals, as the ox, sheep, goat, hair-seal, etc., have been tanned as leather. Leather is used for shoes for the feet of men. In more primitive times it formed a large part of the clothing of men. In Alaska, the stomach and intestines of the sea-lion are used for water-proof rain-coats. It forms a light and serviceable garment. In Japan, slippers are made from the skin of the long-haired native monkey.

The feathers of many birds are used in pillows. Most valuable is the down of the breast of the eider-duck. The most common feathers for pillows come from the hen and the goose. The downy breasts of young eagles are in Alaska stitched together to form mantles. The close-set feathers of the grebe, a diving water-bird, form a kind of fur when the skin is tanned. In Hawaii royal cloaks of great cost have been made, the texture filled and colored with the scarlet and golden feathers of native song-birds.

322. **Animals as ornaments.**—The most valuable ornament derived from any animal is the pearl, the product of a large bivalve mollusk, the pearl-oyster in tropical seas. The pearl is a fine secretion from the mantle of the animal surrounding a grain of sand or other source of irritation. In the museum of Harvard University there is a small fish in-

closed in pearl, it having entered the shell for protection. Coarser pearls occur in other species. Occasionally valuable ones are found in the river-mussel (*Unio*) of our American streams. Other mollusks, as the abalone on the Pacific coast and the *Unio*, furnish material for buttons or for inlaid work. The large scales of certain fishes (as the tarpon) serve a similar purpose. The shells of many small sea-snails are used as beads, and shells of mollusks as well as skins of birds have been used as money by native aboriginal races.

The fine furs above enumerated are largely used for ornament rather than for use. The plumes of various birds, notably the ostrich and the egret, are used for ornament. Dead birds and parts of birds have been worn as ornaments on the hats of unthinking women of even civilized races. Incalculable injury has been done through this fashion by the destruction of insect-eating birds. The singing-birds of Japan have been practically exterminated by the remorseless demands of the milliner, and those of our own country have been greatly reduced in number. Recently our statutes have protected our own singing-birds, but the remorseless destruction of parrots and egrets in Mexico and of terns and other beautiful sea-birds for ornamental purposes still goes on.

323. **Animal products used in the arts.**—Chief among the animal products used in the arts is leather. This is derived chiefly from the skins of the ox, sheep, and goat, but that of the horse, hog, deer, and many other animals, native and domestic, has a certain value. Waterproof shoes are made from the skin of the hair-seal. The skin of the alligator is often tanned for portmanteaus, and the skin of snakes for purses.

Oil is procured from many animals. The fine oil from the liver of the codfish is largely used in medicine, being readily assimilated. Coarser oils are produced from other fishes, especially from the liver of sharks. Still coarser oils are taken in large quantities from the blubber of the right whale and other large whales. These whales also yield whalebone from structures used in straining out their food developed on the roof of the mouth. Finer oils come from the sperm-whale, the most valued being a wax-like substance, spermaceti, found in the head. Ambergris is an abnormal secretion sometimes found about the liver of the sperm-whale. It has a very high value, being very fragrant and used as a perfume. The old-fashioned perfumery musk is a secretion of the musk deer; civet is produced by the Viverra, a weasel-like animal of South America.

Ivory comes from the tusks of the elephant. The long teeth of the walrus are very similar to true ivory.

The horns and bones of different animals have many uses in the arts. The quills of birds have also certain uses. The scales of sea-tortoises are used in making combs. The coloring-matter, cochineal, much used before the introduction of anilin dyes, came from the bodies of certain scale insects living on cactus, chiefly in Mexico.

324. **Animals as enemies and destroyers of enemies of man.**—The chief enemies of man are the larger carnivorous animals which are personally dangerous to him, to his flocks, or to his crops. Among these are the great wildcats (lion, tiger, leopard, panther, etc.), the wild dogs (wolf, coyote, jackal, etc.), the rodents (rabbits, rats, mice) which devour his crops or his stores, and the great multitude of insects which destroy his crops.

Against the great carnivora, the dog, itself a tamed wolf, is his best defender. The dog, cat, ferret, and mongoose have been brought into his service as destroyers of rodents; the mongoose, by destroying lizards and birds' nests also, doing more harm than good. The hawks, owls, and snakes also help him in keeping down the numbers of rats and mice.

Against the hordes of noxious insects man's chief natural protection is found in the singing-birds, most of whom feed on insects which destroy vegetation. He has also a large number of allies among the insects themselves, the helpful forms who destroy those who are noxious or mischievous.

325. Economic entomology.—The enormous number of insects which feed on useful plants gives this branch of science a great practical importance. Most insects feed on plants, and those cultivated by man seem to be especially chosen. This is due to the great masses of the same species brought together in cultivation. An apple orchard of 300 acres in New York gives opportunity for the breeding of enemies of the apple. A 3,300-acre vineyard in California breeds in numbers the insect enemies of the vine. The great orchards and gardens may be compared to bounteous feasts, and the insect guests come in families and remain until the feast is over.

Probably half the insect species inhabiting the United States (upward of 60,000 in all) are injurious to vegetation. There is scarcely a plant, wild or cultivated, that does not harbor insect pests, there being on an average six insect enemies to each species. In Europe 500 species are said to attack the oaks, and 400 species the willows. In America 250 species feed in one way or another on the apple.

Of each species the number of individuals is enormous, for most of them are excessively prolific. Of aphides or plant lice there are 12 generations in a year; 12,000,000 aphides have been found on a single cherry-tree. The grape-destroying insect phylloxera was first discovered in New York in 1854. It was carried to France in 1863, and in 1879 the valuable vines on 3,000,000 acres in France had been destroyed by its root-attacking larvæ.

326. Orchard pests.—In 1878 the "cottony-cushion scale" was brought to California from Australia on an orange-tree. In less than ten years almost every orange-tree in California was attacked by it, and the industry seemed doomed to destruction. The pest was checked by the introduction of its natural enemy, an Australian lady-bird beetle, called

Vedalia, equally prolific and quick to spread from tree to tree.

The so-called San José scale, long known in California, but probably introduced from Asia, is now the worst pest of the orchards of the United States. It is found in 35 States, and in most of these statutes exist aiming at its suppression.

327. Amount of insect destruction.—In 1864 the loss of wheat and corn in Illinois caused by the chinch-bug amounted to $73,000,000. In 1874 the total loss in the United States amounted to $100,000,000. In 1874 the Rocky Mountain locust destroyed in the States between the Missouri River and the Rocky Mountains crops amounting to $100,000,000.

From 1864 to 1878 the cotton-worm in the Southern States destroyed each year $30,000,000 worth of cotton. The Hessian fly has often destroyed $50,000,000 worth of grain in a single year in the United States. Ten per cent. of the field-crops of our country are each year by noxious insects. The annual loss from this source has been calculated to be $300,000,000.

328. Prevention of insect ravages.—These ravages can not be prevented, but they can be materially checked. A few examples of insect fighting may be given:

The cottony-cushion scale was virtually exterminated by the introduction of its enemy at home. This can be done with various other species.

In Indiana the destructive corn-root worm was destroyed by rotation of crops, introducing something on which the larvæ could not feed.

Many insects are killed by insecticides, washes, and sprays, or by fumigation with gas. The Department of Agriculture maintains a division devoted to insect fighting. It costs now about $75,000 a year, and saves the farmers and gardeners many times that amount.

The preservation of insect-eating birds is an effective method of insect fighting.

329. **Beneficial insects.**—Many insects are useful from their habit of devouring the noxious species. The lady-bird beetles feed on scale-insects and plant-lice. The ichneumon flies lay their eggs in the larvæ of many species. The carrion-beetles and others are valuable as scavengers, as are various flies.

330. **Animals as servants.**—As servants of man, the horse, the donkey, the ox, the goat, the dog, the elephant, the camel, the llama, the reindeer, the buffalo of Europe, the water-buffalo of the East Indies, have been with him from the dawn of history, and the help they render needs no description here.

331. **Animals as friends.**—In the category of higher service to man, the service of friendship, the dog stands nearest. The cat always thinks of herself first, but the dog will lay down his life for his master, or even for his own feeling of duty. The monkey is devoted to his own kind, and may be equally devoted to his master, while his thoughts and disposition run in closer parallelism. But the monkeys for the most part are subject to violent fits of passion over which at the best they have little control. The anger or jealousy of some of the larger monkeys is often dangerous to human life. For this reason men have rarely admitted monkeys to their circle of personal friends. In some respects the gentle, wistful little marmosets of South America constitute an exception to the rule of quick temper among monkeys. To the circle of personal intimacy the dog can often rise, and the horse also so far as he can understand.

Other friends of man are the singing-birds, those who can be happy even though caged. Easily first of these is the mocking-bird. The bobolink, most joyous of birds, the nightingale, sweetest of all singers, the wood-thrush, and the skylark can scarcely be reared in cages. Other attractive cage-birds are the cardinal grosbeak, the canary-bird, the Japanese finch, and the many species of parrot, who use their hours of loneliness in human society by picking up and repeating the phrases they hear. They show a skill in imitation and a capacity of association of ideas unequaled by any other of the lower animals. In Japan certain kinds of chirping insects, cicadas and the like, are kept in cages in homes, affording great delight to their owners.

CHAPTER XXVIII

THE ANIMALS OF THE PAST

332. Extinct animals—The mammoth.—The animals alive to-day are but the merest fraction of all that have been. New species of animals long since vanished from the face of the earth are continually being discovered. Notable among fossilized remains are those of the mammoth, an enormous elephant, specimens of which, not even decayed, have been found frozen in the ice of northern Siberia. Some of the specimens discovered were complete with skeleton, flesh, and hair. Its like exists no longer. It resembled an elephant much more than any other kind of living animal, but it was twice the bulk and weight of the largest

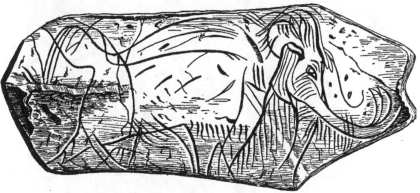

Fig. 242.—Rough drawing of a mammoth, on its own ivory, by a contemporary man.—After Le Conte.

living elephant and a third taller. Its body was covered in places with a brownish wool, in others with long hair. Bones and other remains of many mammoths have been

417

found in various parts of America and Europe. It is be.
lieved that immense herds of this great mammal once

FIG. 243.—The mastodon (*Mastodon americanus*).—After LUCAS, from a painting by J. M. GLEESON.

roamed all over Europe and the northern parts of North
America. But no living mammoths are known. It is an
extinct species of elephant.

There was once found in France a rude drawing (Fig. 242) of the mammoth made on ivory cut from its own tusks, evidently sketched by a man living in that time. This drawing shows that the mammoth was not extinct at the time of the earliest man.

Still another huge extinct animal resembling the elephant, but with very different teeth, is known as the mastodon.

333. Extinct birds.—In New Zealand and Madagascar are found bones and eggs of huge birds (*Dinornis* and *Æpyornis*) which must have been twelve feet high, and which had toe-bones as large as those of an elephant. These birds are long since extinct, and they are not even recorded in history. In Mauritius there once lived a heavy, clumsy bird called the dodo. No living specimens exist, although a few live dodos were known not more than one hundred and fifty or two hundred years ago. It was unable to fly, weighed as much as fifty pounds, and was covered with soft, downy feathers like a new-born chicken. This bird has become extinct in comparatively recent years. Several stuffed specimens are still to be found in museums.

334. Animals becoming extinct.—In New Zealand there may be still living a few individuals of another strange, wingless bird called the *Apteryx*. This bird is disappearing in our own times. Similarly in North America the bison, or buffalo, which roamed the great Western plains in enormous herds only a score of years ago, is now represented by not more than a few hundred individuals living in a state of nature. A few hundred others may be found in zoological gardens and parks. The extinction of the North American buffalo is taking place in our generation. The great auk, a large sea-bird of the Atlantic, has disappeared during the present century. The sea-cow (*Hydrodamalis*), a huge, herbivorous creature, living in the sea and feeding on seaweed, was one hundred and fifty years ago abundant about the Commander Islands, off Kamchatka. It was used as food by sailors, and thus was soon destroyed. It is now known only by its bones preserved in a few museums. The passenger-pigeon of America, which migrated north and south through the Mississippi Valley in flocks of such countless numbers that the sun was darkened as it passed, and which loaded the forest trees of Kentucky with its nests a few decades ago, is now a rare bird, a treasure in the museums.

335. **Fossils.**—Of all these recently extinct animals, we have preserved to us bones, or stuffed specimens, or eggs, as well as the records of personal observation. But we know of the former existence of thousands and thousands of other animals, now extinct, through remains of another kind. These are either actual remains of bones or other parts preserved intact in soil or rocks, or else, and more commonly, parts of the animals which have been turned into stone, or of which stony casts have been made. All such remains buried by natural causes are called fossils. The process by which they are sometimes changed from animal substance into stone is called petrifaction.

Fossils may be of three kinds. In the case of recently extinct animals, bones or other parts of the body may become buried in the soil and lie there for a long time without any change of organic into inorganic matter. Thus fossil insects are found with the bodies preserved intact in amber, a fossil resin from some ancient and extinct pine-tree. Over 800 species of extinct insects are known from amber fossils. The bones of the earliest members of the elephant family, the teeth of extinct sharks, and the shells of extinct mollusks have also been found intact, still composed of their original matter.

In the second kind of fossils the original or organic matter is gone, the organic form and organic structure being preserved in mineral matter. That is, the organic matter has been slowly and exactly replaced by mineral matter. As each particle of organic matter passed away by decay, its place was taken by a particle of mineral matter. These are called petrifactions. This is beautifully shown in the case of petrified wood. We can cut and grind thin a bit of petrified wood, and see in it, with the microscope, the exact details of its original fine cellular structure. This substituted mineral matter may be almost any mineral, but usually it is silica (quartz), or carbonate of lime (limestone), or sulphide of iron (iron pyrites). In the case of animal parts which were originally partly organic and partly inorganic, as bones and teeth and shells, often the organic matter only is replaced by the petrifying mineral, although sometimes the old inorganic matter is also thus replaced. Finally, sometimes the organic matter and organic structure are both lost, only the original outline or form of the whole part being retained. This occurs when the organic matter imbedded in mud and clay decays away, leaving a hollow which is filled up by some mineral different from the matrix. In this case the fossil is simply a cast of the original organic remains.

336. **Fossil-bearing rocks and their origin.**—Examination and study of the rocks of the earth reveal the fact that fossils, or the remains of animals and plants, are found in certain kinds of rocks only. They are not found in lava, because lava comes from volcanoes as a red-hot, viscous liquid, which cools to form the hard lava. No animal or plant caught in a lava stream will leave any trace. Furthermore, fossils are not found in granite, nor in metals, nor in certain other of the common rocks. Many rocks are, like lava, of igneous origin; others, like granite, although not originally in melted condition, have been so heated subsequent to their formation, that any traces of animal or plant remains in them have been obliterated. Fossils are found almost exclusively in rocks which have been formed by the slow deposition in water of sand, clay, mud, or lime. The sediment which is carried into a lake or ocean by the streams opening into it sinks slowly to the

bottom of the lake or ocean and forms there a layer which gradually hardens under pressure to become rock. This is called sedimentary rock, or stratified rock, because it is composed of sediment, and sediment always arranges itself in layers or strata. In sedimentary or stratified rocks fossils are found. The commonest rocks of this sort are limestone, sandstone, and shales. Limestone is formed chiefly of carbonate of lime; sandstone is cemented sand; and shales, or slaty rocks, are formed chiefly of clay.

337. **Sedimentary rocks.**—The formation of sedimentary rocks has been going on since land first rose from the level of the sea; for water has always been wearing away rock and carrying it as sediment into rivers, and rivers have always been carrying the worn-off lime and sand and clay downward to lakes and oceans, at the bottoms of which the particles have been piled up in layers and have formed new rock strata. But geologists have shown that in the course of the earth's history there have been great changes in the position and extent of land and sea. Sea-bottoms have been folded or upheaved to form dry land, while regions once land have sunk and been covered by lakes and seas. Again, through great foldings in the cooling crust of the earth, which resulted in depression at one point and elevation at another, land has become ocean and ocean land. And in the almost unimaginable period of time which has passed since the earth first shrank from its condition of nebulous vapor to be a ball of land covered with water such changes have occurred over and over again. They have, however, all taken place slowly and gradually. The principal seat of great change is in the regions of mountain chains, which, in most cases, are simply the remains of old folds or wrinkles in the crust of the earth.

338. **Deposition of fossils.**—Now we may see how fossils come to exist in the sedimentary rocks. When an aquatic animal dies it sinks to the bottom of the lake or ocean, unless, of course, its flesh is eaten by some other animal.

Even then its hard parts will probably find their way to the bottom. At the bottom the remains will soon be covered by the always dropping sediment. They are on the way to become fossils. Some land animals also might, after death, get carried by a river to the lake or ocean, and find their way to the bottom, where they, too, will become fossils. Or they may die on the banks of the lake or ocean and their bodies may get buried in the soft mud of the shores. Or, again, they are often trodden in the mire about salt springs or submerged in quicksands. It is obvious that aquatic animals are far more likely to be preserved as fossils than land animals. This inference is strikingly proved by fossil remains. Of all the thousands and thousands of kinds of extinct insects, mostly land animals, comparatively few specimens are known as fossils. On the other hand, the shell-bearing mollusks and crustaceans are represented in almost all rock deposits which contain any kind of fossil remains.

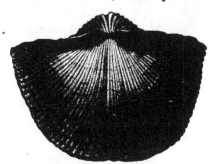

FIG. 244.—A fossil brachiopod (*Spirifer cameratus Morton*). Coal measures, St. Joseph, Mo.

It is obvious that any portion of the earth's surface covered by stratified rocks must have been at some time under water, the bottom of a lake or ocean. If now this portion shows a series of layers or strata of different kinds of sedimentary rocks, it is evident that it must have been under water several times, or at least under different conditions. It is also evident that fossils found in this portion of the earth will contain remains of only those animals which were living at the various times this portion of the earth was under water. Of the animals which lived on it when it was land, there will be no trace, except, possibly, a few land or fresh-water forms which might be swept into the

sea or might be preserved in the mud of small ponds.
That is, instead of finding in the stratified rocks of any
portion of the earth remains of all the animals which have
lived on that portion since the earth began, we shall find, at
best, only remains of a few kinds of those animals which
have lived on this portion of the earth when it was covered
by the ocean or by a great lake.

339. **Geological epochs and their animals.**—Thus, the
great body of fossil remains of animals reveal only a broken
and incomplete history of the animal life of the past.　But

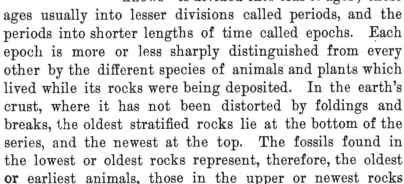

the record, so far as it
goes, is an absolutely
truthful one, and when
the many deposits of fossils in all parts
of the different continents are examined
and compared, it is possible to state numerous general truths in regard to past
life and the succession of animals in time.
The science of extinct life is known as
paleontology.

The study of Paleontology has revealed
much of the history of the earth and its
inhabitants from the first rise of the land
from the sea till the present era.　This
whole stretch of time—how long nobody
knows—is divided into eras or ages; these

Fig. 245 —A Pterodac-
tyl or flying reptile
(*Rhamphorhynchus
gemmingi*), Jurassic
of Bavaria.—After
Zittel.

ages usually into lesser divisions called periods, and the
periods into shorter lengths of time called epochs.　Each
epoch is more or less sharply distinguished from every
other by the different species of animals and plants which
lived while its rocks were being deposited.　In the earth's
crust, where it has not been distorted by foldings and
breaks, the oldest stratified rocks lie at the bottom of the
series, and the newest at the top.　The fossils found in
the lowest or oldest rocks represent, therefore, the oldest
or earliest animals, those in the upper or newest rocks

the newest or latest animals. An examination of a whole
series of strata and their fossils shows that what we call
the most specialized or most highly organized animals did
not exist in the earliest epochs of the earth's history, but

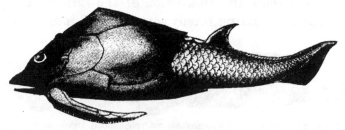

Fig. 246.—An Ostracoderm (*Pterichyodes milleri*), Lower Devonian of Scotland.—
After TRAQUAIR. (The jointed appendage on the head is not a limb.)

that the animals of these epochs were all of the simpler or
lower kinds. For example, in the earlier stratified rocks
there are no fossil remains of the backboned or vertebrate
animals. When the vertebrates do appear, through several
geological epochs they are fishes only, members of the lowest group of backboned animals. More than this, they
represent generalized types of fishes which lack many of

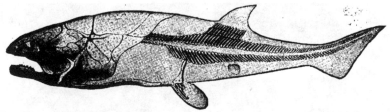

Fig. 247.—An Arthrodire (*Coccosteus decipiens*), Lower Devonian of Scotland.—After
WOODWARD.

the special adaptations to marine life that modern fishes
show. For this reason, they bear a greater resemblance to
the earlier reptiles than do the fishes of to-day. In other
words, they were a generalized type, showing the beginnings of characters of their own and other types. It is
always through generalized types that great classes of
animals approach each other.

In a later epoch the batrachians or amphibians appeared; in a still later period, the reptiles; and last of all, the birds and the mammals, the last being the highest of the back-boned animals. On the opposite page is shown a table giving the names and succession of the various geological periods, and indicating briefly some of the kinds of animals

FIG. 248.—A Crossopterygian fish (*Osteolepis macrolepidotus*). Devonian of Scotland.—From Zittel, after PAUDER.

living in each. In each of these divisions of geological time some one class of animals was especially numerous in species, and was evidently the dominant group of animals through that period. The different ages are therefore spoken of in terms of the prevailing life. Thus, the Silurian Age is known as the age or era of invertebrates; the Devonian, as the age of fishes. In the same way we have the

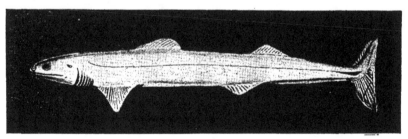

FIG. 249.—*Cladoselache fyleri* (Newberry).—After DEAN, from Devonian rocks in Ohio. The most primitive of known sharks.

Reptilian Age, the Mammalian Age, according to the great class of animals predominating at that time. Of course, in each of the later epochs there lived animals representing the principal classes or groups in all of the preceding ones, as well as the animals of that particular group which may have first appeared in this epoch, or was its dominant group.

Ages.	Eras.	Animals especially characteristic of the age or epoch.
CENOZOIC. AGE OF MAMMALS.	Quaternary or Pleistocene (era of man and insects).	Man; the mammals mostly of species still living.
	Tertiary { Pliocene Miocene Eocene	Mammals abundant; belonging to numerous extinct families and orders.
MESOZOIC. AGE OF REPTILES.	Cretaceous.	Bird-like reptiles; flying reptiles; toothed birds; first snakes; bony fishes.
	Jurassic.	First birds; giant reptiles; ammonites; clams and snails abundant.
	Triassic.	First mammals (a marsupial, lowest kind of mammals).
PALEOZOIC. AGE OF INVERTEBRATES.	Carboniferous (era of amphibians).	Earliest true reptiles. Amphibians. First crayfishes ; insects abundant; spiders; freshwater mussels.
	Devonian (era of fishes).	First amphibian (frog-like animals); cartilaginous fishes, shark-like or mailed; first land shells (snails); shell-fish abundant; first crabs.
	Silurian (era of invertebrates).	First truly terrestrial or airbreathing animals; first insects ; corals abundant; mailed fishes.
	Ordovician or Lower Silurian.	First fishes, probably shark-like with cartilaginous skeleton; brachiopods; trilobites, mollusks, etc.
	Cambrian.	Invertebrates only.
ARCHEAN.	Algonkian. Laurentian.	Simple marine invertebrates.

But certain subdivisions of a principal group or class of animals often appear in an early epoch, become very abundant and highly specialized in a later one, and almost wholly or even totally disappear in a still later one. For example, a group of certain curious animals called Trilobites (Fig. 250), belonging to the great class Crustacea (which includes the crabs, crayfishes, and lobsters), first appeared in

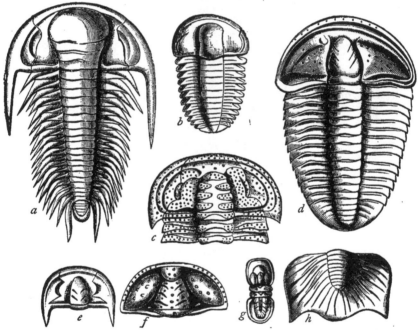

Fig. 250.—Cambrian Trilobites. *a, Paradoxides Bohemicus*, reduced in size; *b, Ellipsocephalus Hoffi; c, Sao hirsuta; d, Conocoryphe Sultzeri* (all the above, together with Fig. *g*, are from the Upper Cambrian or "primordial zone" of Bohemia); *e*, head-shield of *Dikellocephalus Celticus*, from the Lingula flags of Wales; *f*, head-shield of *Conocoryphe Matthewi*, from the Upper Cambrian (Acadian group) of New Brunswick; *g, Agnostus rex*, Bohemia; *h*, tail-shield of *Dikellocephalus Minnesotensis*, from the Upper Cambrian (Potsdam sandstone) of Minnesota.—From Nicholson, after BARRANDE, DAWSON, SALTER, and DALE OWEN.

the Cambrian era, became very abundantly represented in the Silurian era, began to decline in the Devonian, and became extinct in the Carboniferous era. This was not the extinction of a single kind or species of animal, but of a large group of animals represented by thousands of species. Another group with a similar history, traced out wholly by the study of fossils, is that of the sea-lilies or crinoids, radiate animals fixed by a stalk to the bottom, in structure resembling starfishes and sea-urchins. By the abun-

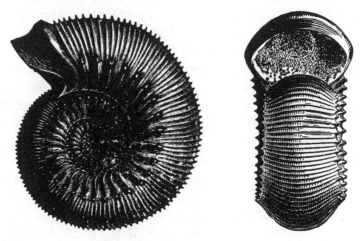

Fig. 251.—Ammonite (*Ammonites humphresianus*) from the Jurassic of Europe.— After Nicholson.

dance and variety of their remains, it is evident that at one time in the earth's history the crinoids were an important and flourishing group of animals. At present, however, there are but very few known living species of crinoids, and these are found only in the unchanging conditions of great depths in the sea. Again, the Nautilus is the only living near relative of what was, in Mesozoic time, a group with hosts of species, the Ammonites (Fig. 251) bearing coiled shells, often very elaborately ornamented.

340. **Man.**—The first traces of man appear in the later geologic epochs in the period called Tertiary. Human bones have been found in caves together with those of the cave-lion, cave-bear, and other extinct animals. In certain lakes in Switzerland and Austria have been found remains of peculiar dwellings, together with ancient fishing-hooks and

a variety of implements of stone and bronze. These houses were built on piles in the lakes, and connected with the shore by long piers or bridges. The extinct race of men who lived in them are known as lake-dwellers. Relics of man, especially rough stone tools and flint arrow- and axe-heads, and skulls and other bones, have been found under circumstances which indicate with certainty that man has existed long on the earth. In Java are found some ancient bones of man-like animals (*Pithecanthropus* and *Anthropithecus*), different, however, from any species or race of men living to-day, and showing traits which indicate a closer relationship with lower animals. The time of historic man—i. e., the period which has elapsed since the history of man can be traced from carvings or buildings or writings made by himself—is short indeed compared with that of prehistoric man. Barbarous man writes no history and leaves no record save his tools and his bones. Iron and bronze rust, bones decay, wood disappears. Only stone implements remain to tell the tale of primitive humanity. These give no exact record of chronology.

So of the actual duration of man's prehistoric existence we can make no estimate. Speaking in terms of the earth's history, man is very recent, the latest of all the animals. In terms of the history of man, he is very ancient. The exact records of human history cover only the smallest fraction of the period of man's existence on earth.

341. Light thrown on zoology by paleontology.—It is plain that much is to be learned, especially in regard to the relationships existing among living animals, by a study of those of the past. A comparison of certain of the ancient reptiles with the long-tailed *Archæopteryx* (Fig. 252) and other toothed birds show that the birds and reptiles were once scarcely distinguishable, although now so very different. Birds have feathers, reptiles do not; and there is scarcely any other permanent resemblance. Fossils show a similar close relation between amphibians and fishes. A

study of these ancient forms also throws light on many
conditions of structure in modern animals otherwise difficult to understand.   For example, while most of the animals closely related to the horse have five toes on each of
their feet, the horse has only one.   We know that the leg

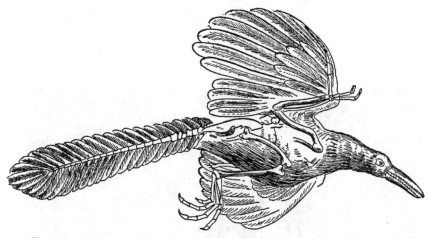

FIG. 252.—Saurian bird with jointed tail, claws on wings, and teeth in jaws (*Archæopteryx lithographica*), from the Jurassic rocks of Bavaria.—After NICHOLSON,
from Owen.

and foot of a horse are homologous with the leg and foot
of a dog, yet a dog's foot has five (on the hind foot usually
four) toes, while the horse's foot has never more than one
toe.   But the study of the ancient horses makes all this
clear.   The remains of over thirty different ancient horse-
like animals have been found in the rocks of the Tertiary
era.   The *Eohippus*, the earliest of these horse-like animals,
found in the oldest Tertiary rocks, was little larger than a
fox, and its fore feet had four hoofed toes, with the rudiment of a fifth, while the hind feet had three hoofed toes
(Fig. 253).

In later rocks is found the *Orohippus*, also small,
but with the rudimentary fifth toe of the fore foot gone.
Still later appeared the *Mesohippus* and *Miohippus*, horses
about the size of sheep, with three hoofed toes only on both

fore feet and hind feet, but with the rudiment of the fourth
toe in the fore feet, of the same size in *Mesohippus*, smaller
in *Miohippus*. Also, the middle toe and hoof of the three
toes in each foot was distinctly larger than the others in
both *Mesohippus* and *Miohippus*. Next came the *Protohippus*, a horse about the size of a donkey, with three toes,
but with the two side toes on each foot reduced in size, and
probably no longer of use in walking. The middle toe and
hoof carried all the weight. Still later in the Tertiary era

Fig. 253.—Extinct four-toed horse (*Eohippus*) from the Eocene of Wyoming, 16
inches high.—After W. D. Matthew, painting by C R. Knight.

lived the *Pliohippus*, an "almost complete horse." The side
toes of *Pliohippus* are reduced to mere rudiments or splints.
This animal differs from the present horse somewhat in
skull, shape of hoof, length of teeth, and other minor details. Lastly came the present horse, *Equus*, with the
splint bones or concealed rudiments of the side toes very

small, and the hoof of the middle toe rounder. In spite of the great difference between the one-toed foot of the living horse and the dog's five-toed foot, there was once a kind of horse which had a five-toed foot, and there is after all a close relationship between the foot of the horse and the foot of the dog.

342. Parallelism of embryonic stages with fossil series.—One of the most important truths of paleontology is that the ancient groups in any type agree more or less closely in structure with the embryos or with the larval stages of the living representatives of the same group. Embryos are generalized organisms simple in structure as compared with the adult animal. The earlier representatives of any class or type

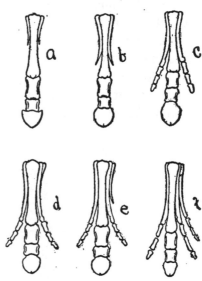

Fig. 254.—Feet in fossil pedigree of horse.—After MARSH. *a, Equus,* Quaternary (recent); *b, Pliohippus,* Pliocene; *c, Protohippus,* Lower Pliocene; *d, Miohippus,* Miocene; *e, Mesohippus,* Lower Miocene; *f, Orohippus,* Eocene.

of animal are likewise simple and devoid of specialization. And there is a curious parallelism, which is not accidental, in the resemblance in the successive stages of animal life in a series of fossils to the successive stages in the embryo of recent forms. The persistence of heredity is undoubtedly the cause of this parallelism. By its influence ancestral traits are repeated in the embryo, even though the characters thus produced give way in later development to further specialization or growth along other lines. This great truth has been stated in these words: "The life-history of the individual is an epitome of the life-history of the group to which it belongs." This statement is only true when stated very broadly, for there are many exceptions or modifications. The embryonic or larval animal is subject to almost endless secondary changes and adaptations whenever these changes are for the advantage of the animal. In general, the simpler the structure of the animal and the less varied its relations in life, the more perfectly are these ancient phases of heredity preserved in the process of development. In such case, the more perfect the parallelism between the development of the individual and the succession of forms in geologic time.

CHAPTER XXIX

GEOGRAPHICAL DISTRIBUTION OF ANIMALS

343. Geographical distribution.—Under the head of distribution we consider the facts of the diffusion of organisms over the surface of the earth, and the laws by which this diffusion is governed.

The geographical distribution of animals is often known as *zoögeography*. In physical geography we may prepare maps of the earth which shall bring into prominence the physical features of its surface. Such maps would show here a sea, here a plateau, here a range of mountains, there a desert, a prairie, a peninsula, or an island. In political geography the maps show the physical features of the earth, as related to the states or powers which claim the allegiance of the people. In zoögeography the realms of the earth are considered in relation to the types or species of animals which inhabit them. Thus a series of maps of the United States could be drawn which would show the gradual disappearance of the buffalo 'before the attacks of man. Another might be drawn which would show the present or past distribution of the polar bear, black bear, and grizzly. Still another might show the original range of the wild hares or rabbits of the United States, the white rabbit of the Northeast, the cotton-tail of the East and South, the jack-rabbit of the plains, the snowshoe rabbit of the Columbia River, the tall jack-rabbit of California, the black rabbits of the islands of Lower California, and the marsh-hare of the South and the water-hare of the canebrakes, and that of all their relatives. Such a

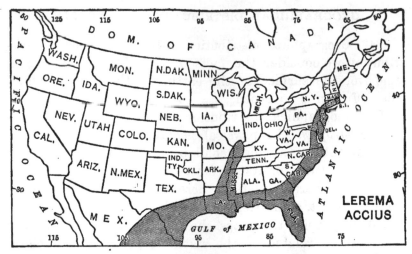

Fig. 255.—Map showing the distribution of the clouded Skipper butterfly (*Lerema accius*) in the United States. The butterfly is found in that part of the country shaded in the map, a warm and moist region.—After SCUDDER.

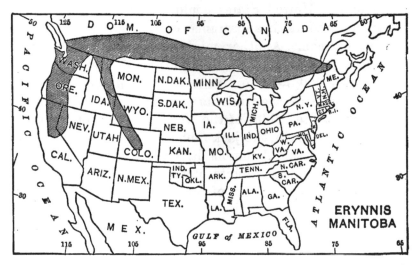

Fig. 256.—Map showing the distribution of the Canadian Skipper butterfly (*Erynnis manitoba*) in the United States. The butterfly is found in that part of the country shaded in the map. This butterfly is subarctic and subalpine in distribution, being found only far north or on high mountains, the two southern projecting parts of its range being in the Rocky Mountains and in the Sierra Nevada Mountains.—After SCUDDER.

436

map is very instructive, and it at once raises a series of questions as to the reasons for each of the facts in geographical distribution, for it is the duty of science to suppose that none of these facts is arbitrary or meaningless. Each fact has some good cause behind it.

344. **Laws of distribution.**—The laws governing the distribution of animals are reducible to three very simple propositions. Every species of animal is found in every part of the earth having conditions suitable for its maintenance, unless—

(*a*) Its individuals have been unable to reach this region, through barriers of some sort; or—

(*b*) Having reached it, the species is unable to maintain itself, through lack of capacity for adaptation, through severity of competition with other forms, or through destructive conditions of environment; or—

(*c*) Having entered and maintained itself, it has become so altered in the process of adaptation as to become a species distinct from the original type.

345. **Species debarred by barriers.**—As examples of the first class we may take the absence of kingbirds or meadowlarks or coyotes in Europe, the absence of the lion and tiger in South America, the absence of the civet-cat in New York, and that of the bobolink or the Chinese flying-fox in California. In each of these cases there is no evident reason why the species in question should not maintain itself if once introduced. The fact that it does not exist is, in general, an evidence that it has never passed the barriers which separate the region in question from its original home.

Local illustrations of the same kind may be found in most mountainous regions. In the Yosemite Valley in California, for example, the trout ascend the Merced River to the base of a vertical fall. They can not rise above this, and so the streams and lakes above this fall are destitute of fish.

346. Species debarred by inability to maintain their ground.
—Examples of the second class are seen in animals that
man has introduced from one country to another. The
nightingale, the starling, and the skylark of Europe have
been repeatedly set free in the United States. But none of
these colonies has long endured, perhaps from lack of adaptation to the climate, more likely from severity of competition with other birds. In other cases the introduced species
has been better, fitted for the conditions of life than the
native forms themselves, and so has gradually crowded out
the latter. Both these cases are illustrated among the rats.
The black rat, first introduced into America from Europe
about 1544, helped crowd out the native rats, while the
brown rat, brought in still later, about 1775, in turn practically exterminated the black rat, its fitness for the conditions of life here being still greater than that of the other
European species.

347. Species altered by adaptation to new conditions.—
Of the third class or species altered in a new environment
examples are numerous, but in most cases the causes in-
volved can only be inferred from their effects. One class
of illustrations may be taken from island faunæ. An island
is set off from the mainland by barriers which species of
land animals can very rarely cross. On an island a few waifs
of wave and storm may maintain themselves, increasing in
numbers so as to occupy the territory; but in so doing
only those will survive that can fit themselves to the new
conditions. Through this process a new species will be
formed, like the parent species in general structure, but
having gained new traits adjusted to the new environment.

To processes of this kind, on a larger or smaller scale,
the variety in the animal life of the globe is very largely
due. Isolation and adaptation give the clew to the formation of a very large proportion of the "new species" in
any group.

FIG. 257.—On the shore of Narborough Island, one of the Galapagos Islands, Pacific Ocean, showing peculiar species of sea-lions, lizards, and cormorants. Drawn from a photograph made by Messrs. SNODGRASS and HELLER

348 **Effect of** barriers.—It will be thus seen that geographical distribution is primarily dependent on barriers or checks to the movement of animals. The obstacles met in the spread of animals determine the limits of the species. Each species broadens its range as far as it can. It

attempts unwittingly, through natural processes of increase, to overcome the obstacles of ocean or river, of mountain or plain, of woodland or prairie or desert, of cold or heat, of lack of food or abundance of enemies—whatever the barriers may be. Were it not for these barriers, each type or species would become cosmopolitan or universal. Man is preeminently a barrier-crossing animal; hence he is found in all regions where human life is possible. The different races of men, however, find checks and barriers entirely similar in nature to those experienced by the lower animals, and the race peculiarities are wholly similar to characters acquired by new species under adaptation to changed conditions. The degree of hindrance offered by any barrier differs with the nature of the species trying to surmount it. That which constitutes an impassable obstacle to one form may be a great aid to another. The river which blocks the monkey or the cat is the highway of the fish or the turtle. The waterfall which limits the ascent of the fish is the chosen home of the ouzel. The mountain barrier which the bobolink or the prairie-dog does not cross may be the center of distribution of the chief hare or the arctic bluebird.

349. **Fauna and faunal areas.**—The term fauna is applied to the animals of any region considered collectively. Thus the fauna of Illinois comprises the entire list of animals found naturally in that State. It includes the aboriginal men, the black bear, the fox, and all its animal life down to the *Amœba*. The relation of the fauna of one region to that of another depends on the ease with which barriers may be crossed. Thus the fauna of Illinois differs little from that of Indiana or Iowa, because the State contains no barriers that animals may not readily pass. On the other hand, the fauna of California or Colorado differs materially from that of adjoining regions, because a mountainous country is full of barriers which obstruct the diffusion of life. Distinctness is in direct proportion to isola-

tion. What is true in this regard of the fauna of any region is likewise true of its individual species. The degree of resemblance among individuals is in strict proportion to the freedom of their movements. Variation within the limits of a species is again proportionate to the barriers which prevent equal and free diffusion.

350. **Realms of animal life.**—The various divisions or *realms* into which the land surface of the earth may be divided on the basis of the character of animal life have their boundary in the obstacles offered to the spread of the average animal. In spite of great inequalities in this regard, we may yet roughly divide the land of the globe into seven principal realms or areas of distribution, each limited by barriers, of which the chief are the presence of the sea and the occurrence of frost. There are the Arctic, North Temperate, South American, Indo-African, Lemurian, Patagonian, and Australian realms. Of these the Australian realm alone is sharply defined. Most of the others are surrounded by a broad fringe of debatable ground that forms a transition to some other zone.

The *Arctic realm* includes all the land area north of the isotherm of 32°. Its southern boundary corresponds closely with the northern limit of trees. The fauna of this region is very homogeneous. It is not rich in species, most of the common types of life of warmer regions being excluded. Among the large animals are the polar bear, the walrus, and certain species of "ice-riding" seals. There are a few species of fishes, mostly trout and sculpins, and a few insects; some of these, as the mosquito, are excessively numerous in individuals. Reptiles are absent from this region and many of its birds migrate southward in the winter, finding in the arctic only their breeding homes. When we consider the distribution of insects and other small animals of wide diffusion we must add to the arctic realm all high mountains of other realms whose summits rise above the timber line. The characteristic large animals of the arctic, as the polar bear or the musk-ox or the reindeer, are not found there, because barriers shut them off. But the flora of the mountain top, even under the equator, may be characteristically arctic, and with the flowers of the north may be found the northern insects on whose presence the flower depends for its fertilization. So far as climate is concerned high altitude is equivalent to high latitude. On certain mountains the different zones of altitude and the corresponding zones of plant and insect life are very sharply defined.

The *North Temperate realm* comprises all the land between the northern limit of trees and the southern limit of frost. It includes, therefore, nearly the whole of Europe, most of Asia, and most of North America. While there are large differences between the fauna of North America and that of Europe and Asia, these differences are of minor importance, and are scarcely greater in any case than the difference between the fauna of California and that of our Atlantic coast. The close union of Alaska with Siberia gives the arctic region an almost continuous land area from Greenland to the westward around to Norway. To the south everywhere in the temperate zone realm the species increase in number and variety, and the differences between the fauna of North America and that of Europe are due in part to the northward extension into the one and the other of types originating in the tropics. Especially is this true of certain of the dominant types of singing birds. The group of wood-warblers, tanagers, American orioles, vireos, mocking-birds, with the fly-catchers and humming-birds so characteristic of our forests, are unrepresented in Europe. All of them are apparently immigrants from the neotropical realm where nearly all of them spend the winter. In the same way central Asia has many immigrants from the Indian realm to the southward. With all these variations there is an essential unity of life over this vast area, and the recognition of North America as a separate (nearctic) realm, which some writers have attempted, seems hardly practicable.

The *Neotropical* or *South American realm* includes South America, the West Indies, the hot coast lands of Mexico, and those parts of Florida and Texas where frost does not occur. Its boundaries through Mexico are not sharply defined, and there is much overlapping of the north temperate realm along its northern limit. Its birds especially range widely through the United States in the summer migrations, and a large part of them find in the North their breeding home. Southward, the broad barrier of the two oceans keeps the South American fauna very distinct from that of Africa or Australia. The neotropical fauna is richest of all in species. The great forests of the Amazon are the dreams of the naturalists. Characteristic types among the larger animals are the snout or broad-nosed (platyrrhine) monkeys, which in many ways are very distinct from the monkeys and apes of the Old World. In many of them the tip of the tail is highly specialized and is used as a hand. The Edentates (armadillos, ant-eaters, etc.) are characteristically South American, and there are many peculiar types of birds, reptiles, fishes, and insects.

The *Indo-African realm* corresponds to the neotropical realm in position. It includes the greater part of Africa, merging gradually northward into the north temperate realm through the transition districts which border the Mediterranean. It includes also Arabia, India, and the neighboring islands, all that part of Asia south of the limit of frost. In monkeys, carnivora, ungulates, and reptiles this region is wonderfully rich. In variety of birds, fishes, and insects the neotropical realm exceeds it. The monkeys of this district are all of the narrow-nosed (catarrhine) type, various forms being much more nearly related to man than is the case with the peculiar monkeys of South America. Some of these (anthropoid apes) have much in common with man, and a primitive man derived from these has been imagined by Haeckel and others. No creature of this character is yet known, but that it may have once existed is not impossible. To this region belong the elephant, the rhinoceros, and the hippopotamus, as well as the lion, tiger, leopard, giraffe, the wild asses, and horses of various species, besides a large number of ruminant animals not found in other parts of the world. It is, in fact, in its lower mammals and reptiles that its most striking distinctive characters are found. In its fish fauna it has very much in common with South America.

The *Lemurian realm* comprises Madagascar alone. It is an isolated division of the Indo-African realm, but the presence of many species of lemur and an unspecialized or primitive type of lemur is held to justify its recognition as a distinct realm. In most other groups of animals the fauna of Madagascar is essentially that of neighboring parts of Africa.

FIG. 258.—A lemur (*Lemur varius*).

The *Patagonian realm* includes the south temperate zone of South America. It has much in common with the neotropical realm from which its fauna is mainly derived, but the presence of frost is a barrier which vast numbers of species can not cross. Beyond the Patagonian realm lies the Antarctic continent. The scanty fauna of this region is little known, and it probably differs from the Patagonian fauna chiefly in the absence of all but the ice-riding species.

The *Australian realm* comprises Australia and the neighboring islands. It is more isolated than any of the others, having been protected by the sea from the invasions of the characteristic animals of the Indo-African and temperate realms. It shows a singular persistence of low or primitive types of vertebrate life, as though in the process of evolution the region had been left a whole geological age behind the others. It is certain that if the closely competing fauna of Africa and India could have been able to invade Australia, the dominant mammals and birds of that region would not have been left as they are now—marsupials and parrots.

It is only when barriers have shut out competition that simple or unspecialized types abound. The larger the land area and the more varied its surface, the greater is the stress of competition and the more specialized are its characteristic forms. As part of this specialization is in the direction of hardiness and power to persist, the species from the large areas, as a whole, are least easy of extermination. The rapid multiplication of rabbits and foxes in Australia, when introduced by the hand of man, shows what might have taken place in this country had not impassable barriers of ocean shut them out.

351. Subordinate realms or provinces.—Each of these great realms may be indefinitely subdivided into provinces and sections, for there is no end to the possibility of analysis. No school district has exactly the same animals or plants as any other, as finally in ultimate analysis we find that no two animals or plants are exactly alike. Shut off one pair of animals from the others of its species, and its descendants will differ from the parent stock. This difference increases with time and with distance so long as the separation is maintained. Hence new species and new fauna or aggregations of species are produced wherever free diffusion is checked by any kind of barrier.

352. **Faunal areas of the sea.**—In like manner, we may divide the oceans into faunal areas or zones, according to the distribution of its animals. For this purpose the fishes probably furnish the best indications, although results very similar are obtained when we consider the mollusks or the crustacea.

The *pelagic* fishes are those which inhabit the open sea, swimming near the surface, and often in great schools. Such forms are mainly confined to the warmer waters. They are for the most part predatory fishes, strong swimmers, and many of the species are found in all warm seas. Most species have special homing waters, to which they repair in the spawning season. To the free-swimming forms of classes of animals lower than fishes, found in the open ocean, the name *Plankton* is applied.

The *bassalian* fauna, or deep-sea fauna, is composed of species inhabiting great depths (2,500 feet to 25,000 feet) in the sea. At a short distance below the surface the change in temperature from day to night is no longer felt. At a still lower depth there is no difference between winter and summer, and still lower none between day and night. The bassalian fishes inhabit a region of great cold and inky darkness. Their bodies are subjected to great pressure, and the conditions of life are practically unvarying. There is therefore among them no migration, no seasonal change, no spawning season fixed by outside conditions, and no need of adaptation to varying environment. As a result, all are uniform indigo-black in color, and all show more or less degeneration in those characters associated with ordinary environment. Their bodies are elongate, from the lack of specialization in the vertebræ. The flesh, being held in place by the great pressure of the water, is soft and fragile. The organs of touch are often highly developed. The eye is either excessively large, as if to catch the slightest ray of light, or else it is undeveloped, as if the fish had abandoned the effort to see. In many cases luminous spots or lanterns are developed by which the fish may see to guide his way in the sea, and in some forms these shining appendages are highly developed. In one form (*Æthoprora*) a luminous body covers the end of the nose, like the headlight of an engine. In another (*Ipnops*) the two eyes themselves are flattened out, covering the whole top of the head, and are luminous in life. Many of these species have excessively large teeth, and some have been known to swallow animals actually larger than themselves. Those which have lantern-like spots have always large eyes.

Fig. 259.—A crinoid (*Rhizocrinus loxotensis*). A deep-sea animal which lives, fixed plant-like, at the bottom of the ocean.

The deep-sea fishes, however fantastic, have all near relatives among the shore forms. Most of them are degenerate representatives of well-known species—for example, of eels, cod, smelt, grenadiers, sculpin, and flounders. The deep-sea crustaceans and mollusks are similarly related to shore forms.

The third great subdivision of marine animals is the *littoral* or shore group, those living in water of moderate depth, never venturing far into the open sea either at the surface or in the depths. This group shades into both the preceding. The individuals of some of the species are excessively local, remaining their life long. in tide pools or coral reefs or piles of rock. Others venture far from home, and might well be classed as pelagic. Still others ascend rivers either to spawn (anadromous, as the salmon, shad, and striped bass), or for purposes of feeding, as the robalo, corvina, and other shore-fishes of the tropics. Some live among rocks alone, some in sea-weed, some on sandy shores, some in the surf, and some only in sheltered lagoons. In all seas there are fishes and other marine animals, and each creature haunts the places for which it is fitted.

(27'

THE END

**THIS BOOK IS DUE ON THE LAST DATE
STAMPED BELOW**

AN INITIAL FINE OF 25 CENTS

WILL BE ASSESSED FOR FAILURE TO RETURN
THIS BOOK ON THE DATE DUE. THE PENALTY
WILL INCREASE TO 50 CENTS ON THE FOURTH
DAY AND TO $1.00 ON THE SEVENTH DAY
OVERDUE.

CPSIA information can be obtained
at www.ICGtesting.com
Printed in the USA
BVHW092212251118
533946BV00001B/51/P